Biomarkers of Arsenic Effects on Gene Expression in Human Skin

The mission of the Awwa Research Foundation (AwwaRF) is to advance the science of water to improve the quality of life. Funded primarily through annual subscription payments from over 1,000 utilities, consulting firms, and manufacturers in North America and abroad, AwwaRF sponsors research on all aspects of drinking water, including supply and resources, treatment, monitoring and analysis, distribution, management, and health effects.

From its headquarters in Denver, Colorado, the AwwaRF staff directs and supports the efforts of over 800 volunteers, who are the heart of the research program. These volunteers, serving on various boards and committees, use their expertise to select and monitor research studies to benefit the entire drinking water community.

Research findings are disseminated through a number of technology transfer activities, including research reports, conferences, videotape summaries, and periodicals.

Biomarkers of Arsenic Effects on Gene Expression in Human Skin

Prepared by:
Jerome Nriagu and **Luda Bernstam**
Department of Environmental & Industrial Health
School of Public Health, University of Michigan
109 Observatory Street, Ann Arbor, MI 48109-2029

Sponsored by:
Awwa Research Foundation
6666 West Quincy Avenue, Denver, CO 80235-3098

U.S. Environmental Protection Agency
Washington D.C.

and

Association of California Water Agencies
Sacramento, CA

Published by:

Awwa Research Foundation
Advancing the Science of Water®

IWA Publishing

DISCLAIMER

This study was funded by the Awwa Research Foundation (AwwaRF). AwwaRF assumes no responsibility for the content of the research study reported in this publication or for the opinions or statements of fact expressed in the report. The mention of trade names for commercial products does not represent or imply the approval or endorsement of AwwaRF. This report is presented solely for informational purposes.

Published by IWA Publishing, Alliance House, 12 Caxton Street, London SW1H 0QS, UK

Telephone: +44 (0) 20 7654 5500; Fax: +44 (0) 20 7654 5555; Email: publications@iwap.co.uk; Web: **www.iwapublishing.com**

AwwaRF report number 91005F. Originally published by AwwaRF for its subscribers in 2004. IWA Publishing version published 2005.

ISBN 1 84339 900 8

CONTENTS

TABLES

FIGURES

FOREWORD

The Awwa Research Foundation is a nonprofit corporation that is dedicated to the implementation of a research effort to help utilities respond to regulatory requirements and traditional high-priority concerns of the industry. The research agenda is developed through a process of consultation with subscribers and drinking water professionals. Under the umbrella of a Strategic Research Plan, the Research Advisory Council prioritizes the suggested projects based upon current and future needs, applicability, and past work; the recommendations are forwarded to the Board of Trustees for final selection. The foundation also sponsors research projects through an unsolicited proposal process; the Collaborative Research, Research Applications, and Tailored Collaboration programs; and various joint research efforts with organizations such as the U.S. Environmental Protection Agency, the U.S. Bureau of Reclamation, and the Association of California Water Agencies.

This publication is a result of one of these sponsored studies, and it is hoped that its findings will be applied in communities throughout the world. The following report serves not only as a means of communicating the results of the water industry's centralized research program but also as a tool to enlist the further support of the nonmember utilities and individuals.

Projects are managed closely from their inception to the final report by the foundation's staff and large cadre of volunteers who willingly contribute their time and expertise. The foundation serves a planning and management function and awards contracts to other institutions such as water utilities, universities, and engineering firms. The funding for this research effort comes primarily from the Subscription Program, through which water utilities subscribe to the research program and make an annual payment proportionate to the volume of water they deliver and consultants and manufacturers subscribe based on their annual billings. The program offers a cost-effective and fair method for funding research in the public interest.

A broad spectrum of water supply issues is addressed by the foundation's research agenda: resources, treatment and operations, distribution and storage, water quality and analysis, toxicology, economics, and management. The ultimate purpose of the coordinated effort is to assist water suppliers to provide the highest possible quality of water economically and reliably. The true benefits are realized when the results are implemented at the utility level. The foundation's trustees are pleased to offer this publication as a contribution toward that end.

Elevated levels of arsenic have been reported in water supplies in many regions of the United States including Oregon, California, Alaska, Michigan, Ohio, Wisconsin, Illinois, and New Hampshire. Epidemiological studies show that the levels of arsenic found in the drinking water in many countries are associated with skin diseases (including skin cancer) as well as internal cancers of the bladder, liver, lung and kidney. It is generally presumed that ingestion of arsenic is the primary exposure route, but no clear pathway has been established for the transport of arsenic from the gastro-intestinal (GI) track to the skin. Arsenic in water, soil and dust readily come into contact with human skin during daily activities, work or play and has the potential to be absorbed from the water and soil matrices. The effect of dermal exposure to elevated levels of arsenic in water (while showering, bathing and hand and face washing) on the development of skin cancer and other skin diseases has not been investigated in a meaningful manner, and is the goal of this study.

Edmund G. Archuleta, P.E.
Chair, Board of Trustees
Awwa Research Foundation

James F. Manwaring, P.E.
Executive Director
Awwa Research Foundation

ACKNOWLEDGMENTS

The authors of this report would like to thank the following colleagues at the University of Michigan who served generously as consultants to the project: Dr. Thomas Glover (Dept of Human Genetics), Prof. Robert Gray (Senior Associate Dean, School of Public Health), Dr. Frizell Vaughan (Associate Professor Emeritus, Department of Environmental Health Sciences, Toxicology Program), Dr. Martin Philbert (Associate Professor, Department of Environmental Health Sciences, Toxicology Program). Inputs from these expert consultants helped us to sharpen the proposal, bring focus to the study and improve the interpretation of the data.

The authors are indebted to the post-doctoral fellow (Dr. Cheng-Hang Lan) and graduate students (Dr. Janice Lee and Ms. Yunseon Chong) who conducted some of the experiments. Their dedication and hard work were instrumental in completing the progress reports on schedule. Our gratitude also goes to Ms. Rhonda Little for skillful preparation of paraffin and electron microscopy sections.

Gratitude is expressed to the following members of Project Advisory Committee for their timely comments and suggestions which have improved the research effort immeasurably: Prof. Curtis Klaassen (Dept. of Pharmacology, Toxicology and Therapeutics, University of Kansas, Kansas City, KS); Prof. Mariano Cebrian (Dept. de Farmacologyia y Toxicologyia Inst. Politecnico Nacional, Mexico); Dr. Mike Hughes (USEPA, National Health and Environmental Effects Research Laboratory, Research Triangle Park, NC); Ms. Krista Clark (ACWA, Sacramento, CA) and Dr. Pankaj Parekh (Los Angeles Department of Water and Power, Los Angeles, CA). Appreciation is extended to project managers Jarmila Popovicova, Frank Blaha and Alice E. Fulmer for helping us to move the project along.

EXECUTIVE SUMMARY

BACKGROUND

The pronounced affinity of arsenic for keratinizing structures makes the skin a critical organ in arsenic toxicity. Systemic toxicity of arsenic is sometimes manifested in a variety of symptoms including keratosis, hyperpigmentation, skin eruptions, dermal lesions (such as arsenical epithelioma, verrucous hyperkeratosis, warts and melanosis), Bowen's disease (precancerous dermatitis), various forms of skin cancer, and neoplasia (Guy et al., 1999). Chronic dermal accumulation of arsenic increases the susceptibility of the skin to ultraviolet radiation. It is generally presumed that ingestion of drinking water is the primary exposure route for arsenic, but no clear pathway has been established for the transport of arsenic from the GI tract to the skin. Arsenic in water, soil and dust come into contact with human skin during daily activities, work or play and has the potential to be absorbed from the water and soil matrices. Absorption of arsenic through the skin has generally been ignored in previous studies on the human exposure to arsenic in the environment.

APPROACH AND RESULTS

This project has examined the effects of As exposure on human skin at four levels of functional cellular organization: morphological and ultrastructural assessment of the damage (gross and microscopic changes, loss of selective permeability), generalized biosynthesis processes (DNA, RNA, protein synthesis), activation of selected stress-induced signal transduction components (HSP70 and p38) and genotoxic effects reflected in DNA fragmentation (TUNEL, COMET, Telomerase assays and expression of p53 protein). The later part of the study was concentrated on the analysis of effects of As on the expression of signal transduction pathway and chemical inhibition of the toxic effects of arsenic. The studies were based mainly on lab-grown human skins which allowed us to achieve a higher degree of standardization than could be expected from random *ex vivo* human skin biopsy material. Although there are some differences between the natural skin and any culture system (for instance, epidermal desquamation is unlikely to occur in culture as is often the case *in vivo*, and there is no active vascular system in a skin culture), the artificial skins offer several unique features for experimentation. For instance, a toxic agent can either be applied topically on the surface of the culture or added to the culture growth medium to reach the basal cells from substratum to mimic the two routes of entry of toxic compounds into the skin system (topically via showering, bathing and hand-washing versus reaching the basal cells from beneath, resembling the *in vivo* route of entry after ingestion of contaminated water and/or food).

Permeation and Retention of Arsenic by Human Skin

This project demonstrates the remarkable affinity of cultured skin cells for both inorganic arsenic in the form of pentavalent arsenic (As(V) or arsenate) and trivalent arsenic (As(III) or arsenite). The absorption rate for arsenic is estimated to be about 6-8%/hr of the applied dose for As(III) and 1-8%/hr of the applied dose for As(V). Easier penetration and binding apparently is related to the higher toxicity of As(III) compared to As(V). Average permeability constant (K_p) was estimated to be 4.3×10^{-5} cm/hr for As(V) and 10×10^{-5} cm/hr for As(III). Back of the envelope calculations suggest that absorption through the skin surface of an adult showering with water containing 100 μg/L As(III) can account for about 20% of the recommended reference dose [0.3 μg/(kg body wt/day)]. The dermal route can account for over 50% of the reference dose if the shower water contains over 500 μg/L of either As(V) or As(III) for both adults and infants. These calculations are consistent with epidemiological studies in Bangladesh which show that skin lesions and vascular diseases are most prevalent in areas where total arsenic concentration in drinking water is 100 μg/L and higher.

Morphological Effects

The effects of absorbed arsenic on structure and integrity of the skin layer are important issues in dermal toxicity. Photomicrographs of exposed skins show that at low doses of arsenic, the ultrastructural changes in the skin included development of thick and compact cornified layers, enlarged nuclei with margination of nuclear chromatin. Prominent keratohyalin-like granules known to contain high molecular weight proteins could be seen indicating that the process of terminal differentiation of keratinocytes was being disturbed. Higher arsenic doses or increased exposure time can result in a reduction in the amount of cornified cell layers, disorganization of nuclear chromatin, appearance of high number of degenerated mitochondria, thickening and shortening of the tonofilaments. At higher doses (up to 1000 μg/L) the progressive nuclear deterioration and largely erased cellular structure made reliable ultrastructural assessment impossible. The appearance of the submerged skin cultures after exposure to arsenic closely resembled the cancerous human skins sometimes described as the "snake skin" appearance. It is not often that one is able to replicate the appearance of a diseased organ in the laboratory.

Cytotoxicity of Arsenic

The release of lactate dehydrogenase (LDH) enzyme from skin cells into the test medium was used to assess the effect of arsenic exposure on cellular membrane integrity. A dose-dependent release of LDH from human keratinocytes (HK) was observed down to concentrations as low as 1.0 μg/L. The MTT assay was used to measure the cell viability after exposure to arsenic. With As(V), no change in cell viability was observed in the concentration range of 10-1000 μg/L. With As(III), however, significant reduction in cell viability (relative to control) was observed at concentrations of 10 μg/L.

The effects of arsenic exposure on DNA, RNA and protein synthesis were investigated using the incorporation of ^{3}H and ^{14}C -labeled compounds. Exposure of submerged HK to 10 μg/L As(III) reduced the extractable protein by 10-25% relative to control. Topical application of 10 μg/L As(III) to skin cultures resulted in a significant reduction of DNA synthesis (thymidine incorporation). When exposure was achieved by adding the arsenic to the growth medium of artificial skin, the DNA synthesis was reduced by 69% with As(III) dose of 10 μg/L, by 95% with 100 μg/L and by over 98% with 1000 μg/L. RNA synthesis does not appear to be particularly sensitive to arsenic exposure—only mild effect was observed on this process even at the 1000 μg/L dose.

Programmed cell death (apoptosis) in human keratinocytes, measured by DNA fragmentation or TUNEL assay, increased progressively with arsenic dose and exposure time with effects observed at As(III) concentrations as low as 10 μg/L. The changes ranged from occasional apoptotic changes (typically after one-hour exposure) to necrotic destruction of cells to complete destruction of the cell structure.

Cells damaged by arsenic develop comet-like tails in gel electrophoresis and the extent of the damage can be estimated by measuring the length, shape and intensity of the "tail" relative to the "head" of the comet. Exposure of human fibroblasts (HF) to 5 μg/L As(III) for only 30 minutes increased the comet moment by 165%, implying an adverse effect dose that was well below 5 μg/L. Changes in the comet moment increased progressively to 195% at 10 μg/L, 282% at 100 μg/L and 728% at 500 μg/L. This study shows that (a) the Comet assay can provide a quantitative measure of cell damage at low doses of arsenic exposure and (b) at a given dose, the adverse effect of As(III) on the HF is more than that on HK.

Effect of arsenic on the chromosomes was evaluated using the Telomerase PCR ELISA method. Enhancement of telomerase activity in HF was observed with exposures to 100 μg/L of As(III) for only 15 and 30 min. Chromosome damage does not appear to be a particularly sensitive biomarker of exposure to low environmental levels of arsenic in human fibroblasts at such short time intervals. It would be desirable to run experiments using low doses (5-10 μg/L) of As(III) for longer time periods to determine whether chromosomal damage can be detected.

Genotoxicity and Signal Transduction Experiments

Marked dose- and time-dependent increase in the tumor suppressor gene p53 expression was noticed in basal and spinous cell layers of artificial human skin exposed to 100 μg/L for 48 hour and to 500 μg/L for 6 hours. Similar activation of p53 gene expression have been reported with various toxic agents and in carcinogenesis and its activation does not appear to agent-specific.

Western blot technique was used to measure changes in the expression of heat shock protein 70 (HSP70) and signal transduction protein 38 (p38) in HK exposed to As(III). Changes in the expressions of both proteins were observed at an As(III) concentration of 10 μg/L, which then increased in a dose-dependent manner.

Exposure of both HK and HF to arsenic even at doses of 5 μg/L for 1-3 hrs produced activation of early stress-related and mitogen-activated kinases. The response of cells was tested in cultures grown submerged and in artificial skin. Submerged cultures showed more intense staining at 5 μg/L than artificial skin.

Comparison of the degree of expression of p38 and JNK in HF and HK suggests that the response of mitogen-activated protein kinase signal transduction pathway in HF is somewhat stronger and detectable at lower As(III) doses that in HK. These are initial observations that need to be confirmed in more extensive and detailed experiments. However, the greater sensitivity of HF to As(III) compared to HK was evident in the other experiments as well as in DNA fragmentation studies by the Comet assay. In both submerged cultures and artificial skin, the sensitivity of cells to arsenite was similar in demonstrating significant activation of p38, ERK and JNK at the low arsenate levels (<5 μg/L) and exposure times as short as 1-3 hrs. The extent of activation of these stress markers was increased in a dose-dependent manner so that progressively higher number of cells showed positive staining. The dose-dependent activation of these signal transduction components was correlated with the progressively enhanced cyto- and genotoxicity of higher doses of As(III) revealed by other biomarkers reported in this study.

Inhibitor Analysis and Prevention of Arsenic Toxicity

The signal cascade triggered by arsenite induced the activity of the mitogen-activated protein (MAP) kinases such as extracellular regulated kinase (ERK), c-jun terminal kinase (JNK) and p38. Through the JNK and p38 pathways, arsenite activated the immediate early genes c-fos, c-jun and egr-1, which are usually activated by various growth factors, cytokines, differentiation signals and DNA damaging agents. This study demonstrated that the effects of arsenite may resemble heat shock stress by being manifested in human keratinocytes as dose-dependent changes in the expression of the signaling protein p38, ERK and JNK and heat shock proteins of the HSP70 family.

Since activation of cyclooxygenase-2 (COX-2) is frequently observed in various cells under stress as well as in malignant cells and is often associated with apoptotic changes in such cells, we conducted an inhibitor study to determine the effect of abrogated activation of COX-2 on the development of arsenic-related toxic damage. We found that COX-2 inhibitor Niflumic acid provided virtually no protection against morphological damage produced at any As(III) concentration tested. The important observation made in these experiments is that the COX-2 inhibitors tested (NS-398, Niflumic acid and Nimesulide) afforded lesser protection against the destructive effects of As(III) than the protection provided by any of the three MAPK inhibitors tested—SB 203580, PD 98059 and SB 202190. This is manifested both at the level of expression of p38 and JNK activation as well as at the level of morphological integrity in As(III)-exposed HF. Thus, it appears that in order to counteract the destructive cytotoxic and genotoxic effects of exposure to As(III), the application of specific inhibitors that can interrupt the MAPK signal transduction pathway is more efficient than the disruption of reactive prostaglandin synthesis induced by arsenic exposure. This leads to the suggestion that a novel protective and preventive approach to chronic and acute As toxicity can be developed using specific inhibition of the signal transduction pathway. Details of specific components in this pathway that can serve as the most appropriate targets for such inhibitor action will have to be studied further.

The preliminary experiments point to the fact that specific inhibition of signal transduction pathway may offer practical benefits in preventing cytotoxic and genotoxic effects of acute and chronic As exposure in susceptible populations.

CONCLUSIONS

Various test methods used in this study show that cultured human skin cells suffer damage (especially at the cellular level) when exposed to As(III) concentrations of 10 μg/L (effects were occasionally observed at 5 μg/L dose) for short periods of time (1-3 hrs). At the molecular level, the activation of early stress-related and mitogen-activated kinases and DNA fragmentation was detected by the Comet assay. At the cellular level, the release of LDH could be observed at As(III) concentrations as low as 1 μg/L, indicating significant damage to the cell membranes.

Alterations in the genomic functions of cells exposed to arsenic could be identified at the chromosome level from DNA fragmentation which was confirmed by the TUNEL assay as well as by electron microscopy (EM). Significant structural damage was confirmed using Comet assay at As(III) dose of 5 μg/L while functional abnormalities were demonstrated by Telomerase–TRAP assay in cells exposed to 100 μg/L As(III) for only 15 min. These abnormalities can be used as potential molecular biomarkers of chronic exposure to low levels of arsenic for the dermal system.

The results also provided new insights for further analysis of the mechanisms of repair of As-damaged cellular functions.

The study presents a novel approach that can be used to prevent or protect against chronic and/or acute As toxicity based on specific inhibition of the signal transduction pathway. The protective or preventive process can be moderated using selective COX-2 inhibitors of the reactive branch of the cellular stress response mechanism represented by activation of the inducible prostaglandin synthesis using COX-2 inhibitors. Details of the specific components in this pathway that can serve as the most appropriate targets for such inhibitor action need further investigation.

CHAPTER 1
INTRODUCTION AND LITERATURE REVIEW

The development of skin cancer has been repeatedly associated with arsenical poisoning in communities in many parts of the world (including Zimbabwe, Taiwan, China, Thailand, Viet Nam, Singapore, India, Bangladesh, Japan, Argentina, Chile and Mexico) with elevated levels of arsenic in their drinking water (Nriagu, 1994; Hopenhayn-Rich et al., 1996; Smith et al., 2000). Arsenic also causes a variety of benign skin lesions including hyperpigmentation and hyperkeratosis in many of these communities (Chen, 1985; Jaafar et al., 1993). In the United States, skin cancer accounts for over one million physician visits annually in the U.S., and the risk of patients developing new primary non-melanoma skin cancer within five years has been estimated to be 50% (Karagas et al., 1992). Non-melanoma skin cancer (NMSC) affects over 600,000 persons annually and is one of the most common forms of cancer in the U.S. (Karagas et al., 1992). Although elevated levels of arsenic have been reported in water supplies in many regions of the United States including Oregon, California, Alaska, Michigan, Ohio, Wisconsin, Illinois, and New Hampshire, a significant correlation between the prevalence of skin diseases and arsenic concentrations in drinking water has not been demonstrated in the country (Smith et al., 1992; NAS, 1999). In addition to arsenic (possibly), other major risk factors for the high incidence of skin cancer in the U.S. population include exposure to UV light, ionizing radiation, and long-term chemotherapy (Karagas et al., 1992) but the relative contributions of these factors to the burden of skin diseases are unknown.

Despite reported cases of cancer development at long-term, low level exposure to arsenic in drinking water (down to 1.2 μg/L, see Wagner et al., 1979), much remains to be established regarding the doses and exposure times necessary for carcinogenicity and cutaneous toxicity (Zaldivar, Prunes, and Ghai, 1981; Shannon and Strayer, 1989). Other factors (such as chronic liver disease, nutritional status and UV exposure) in the development of arsenic-related carcinogencity are recognized confounders (Everall and Dowd, 1978; Hsueh et al., 1995). Skin cancer lesions related to arsenic exposure may also act as early warning signals of the subsequent development of cancers of internal organs that can become manifested many years later (Cowlishaw et al., 1979; Upshaw, Bryant, and Claiborne, 1979; Roat et al., 1982; Tello, 1988; Bates, Smith, and Hopenhayn-Rich, 1992; Tsuda et al., 1995). Human susceptibility to arsenic-induced cancers, combined with the absence of animal models for arsenic carcinogenesis and lack of understanding of the mechanisms of carcinogenic action have provided major challenges in assessing the risks of exposure to arsenic in drinking water in the United States (Goering et al., 1999; NAS, 1999). Although the effects of acute toxicity of arsenic have received much attention, the chronic effects of low-level exposures have been much less understood since they may not be manifested in overt disease in the affected system or population for a long time (Karjalainen, Salo, and Teppo, 1989; Goering et al., 1999). Available evidence points to significant genotoxic effects of arsenic compounds manifested at exposure concentrations frequently found in the environment (Lee and Ho, 1994; Germolec et al., 1996).

An objective of this study is to evaluate the differences in the uptake and effects of arsenate [As(V)] and arsenite [As(III)] on excised or artificial human skin system using methods developed in our laboratory (Bernstam, Vaughan, and Bernstein, 1990; Vaughan, Gray, and Bernstein, 1986; Bernstam, Vaughan, and Bernstein, 1986; Lee and Ho, 1994). Arsenate is believed to gain entry into cells through the phosphate channels and tends to inhibit, among other things, enzymatic reactions in which phosphate esters are formed (Wang et al., 1996). Arsenite, on the other hand, diffuses passively into cells and tends to exert its toxic effects by reacting with molecules containing spatially proximate thiols (NAS, 1999). There is currently little data on differential uptake and retention of As(III) and As(V) by the skin or the metabolism of the arsenic in the skin cells. Although most toxicological studies have focused on the more toxic As(III), it is As(V) that predominates in nature (Nriagu, 1994). Understanding the differential uptake and toxic mechanisms of As(V) and As(III) in the human skin system is believed to be critically important in assessing the risks of exposure to arsenic in our water supply.

Cytoskeletal abnormalities frequently accompany characteristic chromosomal aberration leading to or associated with malignancy. So far, little work has been done on effects of arsenic on the cytoskeleton in keratinocytes except for one study that related an increase in TGF alpha level to sodium arsenite exposure (Ju et al., 1993). The effects of arsenic compounds on the structure and integrity of the cytoskeleton (microfilaments and microtubules) reflect a disruption of microtubule assembly and spindle formation that can be linked to the induction of aneuploid cells (Ramirez et al., 1997). Examination of the spindle apparatus using an anti-beta-tubulin antibody suggests that arsenic may disrupt spindle formation by inhibiting microtubule assembly and inducing tubulin depolymerization (Ramirez et al., 1997). One of the objectives of this study is to evaluate the arsenic-induced changes in morphological characteristics of artificial skin and the integrity of the cell membrane.

The intracellular effects of arsenic that may ultimately result in programmed cell death (apoptosis) include changes in inducible gene expression, inhibition of specific enzyme activity, decreased ATP production, protein phosphorylation, and generation of oxidative stress. Other reported effects of arsenite include: (i) inhibition of the cell cycle progression, (ii) induction of chromatid-type aberrations and sister-chromatid exchanges (SCEs), and (iii) potentiation of the X-ray and UV induced chromosomal damage (Jha et al., 1992) suggesting that sodium arsenite interferes with the DNA repair process. A goal of this project is to evaluate the cytotoxicity of arsenic using various complementary scientific probes.

Oxidative stress produced by different agents (heavy metals, heat shock, photo energy, etc) can trigger a fundamental biological reaction—a heat shock response (HSR) (e.g., Maytin, 1992; Jurivich et al., 1995; Cullen and Sarge, 1997; Schett et al., 1998; Kültz and Csonka, 1999; Miyakawa et al., 1999). The response of cells of different origin to arsenic compounds is, in many respects, similar to their response to other stresses, notably the heat stress. Therefore, the general features of the control of gene expression in arsenic-stressed cells are common with those characteristics of the heat shock response that can be elicited in a wide range of cell types and by a wide range of agents, from temperature to osmotic stress to mechanical impact and even restriction of motion. Like heat shock, arsenic is capable of inducing HSPs of various sizes from the smaller proteins to the larger HSP species (see p. 11). New heat shock proteins are being continually identified in cells exposed to a variety of stress agents. A goal of this project is to investigate, for the first time, the use of HSPs (specifically, HSP70 and HSP53) as potential biomarkers of arsenic exposure.

MECHANISMS OF ARSENIC TOXICITY: AN OVERVIEW

Arsenic is ubiquitous in the environment and is found in small quantities in all rocks, soil, water, air and foods. Human beings are therefore exposed to low doses of arsenic with estimated daily intake from natural sources typically being in the range of 12-40 μg/day (Thornton, 1996).

Since some crops and organisms bioaccumulate arsenic, regional and individual eating habits greatly affect the dietary arsenic intake. For instance, saltwater cod, crabs, lobster, and shrimp typically contain 10-40 μg/g wet weight of arsenic whereas freshwater fish (including pickerel, catfish and coho salmon), pork and beef generally contain <1.0 μg/g arsenic (NAS, 1999). Organic forms of arsenic in seafood tend to be rapidly excreted unchanged (USEPA, 1996) and the health significance of dietary exposure must be viewed in circumspect manner. Studies to date do not provide evidence that arsenic at low doses is an essential element in humans or that it is required for any essential biochemical function (NAS, 1999). At high doses of supplementation, arsenic has been shown to have a growth-stimulating effect on rats, chicken, goats, and guinea pigs (NAS, 1999)

Arsenic contamination of the environment has arisen because of mining and smelting activities and the many industrial applications of arsenic. Today, there is growing concern about the possible health effects of exposure to arsenic from (a) industrial point sources including power stations, metal smelting and processing plants, tanneries and wood treatment plants, (b) lumber treated with chromated copper arsenate and (c) past pesticidal application of lead and other metal arsenates to agricultural soils. The concern is heightened by recent reports of elevated levels of arsenic in groundwater derived from natural sources and the publicity surrounding modification of the natural groundwater cycle which has put millions of people in Bangladesh and West Bengal at risk of being poisoned by arsenic (Ahmad et al., 1997). In response to public concern, many countries and international agencies are lowering their guideline level for arsenic in drinking water (from 50 μg/L to 10 μg/L by the WHO (1993), for example) as well as the maximum allowable daily intake for arsenic (NAS, 1999). The lowering of criteria levels for arsenic in drinking water is being contested on the grounds that it is not supported by sound science.

The absorption and toxicity of arsenic depend on the form in which it is ingested (Leonard, 1991). Soluble inorganic species are most readily absorbed from the gastrointestinal tract, typical absorption rate being 40-100% of the ingested amount (Pontius, Brown, and Chen, 1994). Once absorbed, the biotransformation of arsenic follows several pathways. For methylation to occur, As(V) must first be reduced to As(III) through a process mediated by glutathione (GSH). The arsenite is then methylated to MMA and DMA, primarily in the liver, through methyltransferase enzymes with S-adenosyl-methionine (SAM) as the methyl-donating cofactor (Aposhian et al., 1997). The role of arsenic methylation is enigmatic. It was previously regarded as a detoxification mechanism because the methylated compounds are less toxic, less reactive with biological molecules and are excreted much faster than the inorganic species (Buchet and Lauwerys, 1985; Chen et al., 2003). However, some mammals do not methylate arsenic suggesting that methylation is not the only, or even a primary, detoxification pathway. On the other hand, there is some evidence to suggest that methylated arsenic metabolites contribute to this element's carcinogenicity and genotoxicity (Goering et al., 1999). Furthermore, recent studies that methylation may increase rather than decrease the toxicity of arsenic. The methylation of inorganic arsenic produces short-lived trivalent forms of arsenic that appear to inhibit enzyme activity, increase cell toxicity and have genotoxic properties that may contribute to cancer development (Kitchin, 2001; Yu et al., 2000). Other studies show that arsenic compounds may be effective in cancer chemotherapy directed against some types of acute leukemia (Chen et al., 1996). The mechanisms involved in arsenic toxicity are thus clouded by the fact that arsenic can have both carcinogenic and anticarcinogenic effects.

Bacterial species are known to be resistant to arsenic. Specific plasmids and arsenic resistance operons account for the resistance mechanism (e.g., Crameri et al., 1997; Suzuki et al.,

1997; Cai et al., 1998; Silver, 1998; Xu et al., 1998) and some bacteria use arsenic as substrate for growth (e.g., Newman et al., 1997). The arsenic resistance system in bacteria protects the cells by transporting arsenite [As(III)] with the ArsB polypeptide functioning as a chemiosmotic efflux transporter or with two polypeptides, ArsB and ArsA, functioning as an ATPase (Silver, 1998). The third protein of the arsenic resistance system in bacteria is an enzyme that reduces intracellular arsenate [As(V)] to arsenite [As(III)], the substrate of the efflux system (Silver, 1998).

In contrast, arsenic compounds produce a variety of stress responses, metabolic derangement, growth inhibition and eventually apoptosis in mammalian cells (Eguchi, Kuroda, and Endo, 1997). Moreover, a dramatic difference exists in the sensitivity of human and rodent cells to arsenic stress. In spite of the fact that trivalent arsenic induces DNA-protein crosslinks and appears to exert genotoxic effects by inhibiting repair of DNA damage, various rodent cell lines such as Chinese hamster V79 cells and their arsenite-resistant variants are reported to have an arsenite- and antimonite-inducible tolerance mechanism that protects against the subsequent cytotoxic effects of arsenite, arsenate and antimonite (Gebel et al., 1998). However, numerous attempts to develop arsenic tolerant variants of human normal diploid cells or human tumor cells of different origins failed to elicit an inducible tolerance response to arsenite (Rossman et al., 1997). Moreover, all the human cell lines tested proved to be much more sensitive to arsenite than Chinese hamster cells, with human keratinocytes being especially sensitive. This helps to explain why rodents do not develop tumors as a result of exposure to arsenic and humans do (Rossman et al., 1997).

In exposed human populations, arsenic is associated with tumors of the skin, bladder, liver and lung (NAS, 1999). The mode of action of arsenic in carcinogenesis has not been established, due to the lack of validated, reproducible animal cancer models—humans are more sensitive than experimental animals to cancers induced by arsenic as noted above. Critical inferences from recent studies include: (a) arsenic-initiated oxidative stress has been linked to chromosomal abnormalities; (b) arsenic-produced alteration in DNA repair has been found to interfere with cell proliferation; (c) none of the inorganic arsenic species has been shown to interact with DNA as point mutagens and are not DNA-reactive electrophiles; (d) inorganic arsenic produces chromosomal-type damage and can have a profound influence on cell differentiation (Chen et al., 1996; Goering et al., 1999). Various mechanisms by which arsenic may induce cancer at the cellular/molecular level have been proposed, but no unified hypothesis has emerged (see NAS, 1999; Aposhian et al., 1997) that can account for all the observations on the enigmatic behavior of arsenic.

Morphological and Cytoskeletal Effects

The effects of arsenic compounds on the structure and integrity of the cytoskeleton are poorly understood. Studies on the cytoskeletal integrity are of particular interest for evaluating carcinogenic potential of compounds since cytoskeletal abnormalities frequently accompany characteristic chromosomal aberrations leading to or associated with malignancy. Morphological abnormalities in cells exposed to various arsenic compounds are striking and suggest underlying disruption of intracellular structural systems responsible for maintaining cellular morphology, shape and locomotion.

The Skin

The association of arsenical poisoning with the development of hyperpigmentation, keratoses and skin cancer has been repeatedly documented in chronic arsenical poisoning from drinking water contaminated with arsenic (e.g., Jaafar et al., 1993; Kovacs et al., 1996; Ohnishi et al., 1997). It appears that skin cancer lesions related to arsenic exposure act as early warning signals for subsequent development of cancers of internal organs many years later (Tello, 1988; Tsuda et al., 1995; Wong, Tan, and Goh, 1998). Some hyperkeratotic lesions and squamous cell carcinomas in situ may progress to invasive carcinoma; other invasive squamous cell carcinomas will develop de novo. These cutaneous squamous cancers may metastasize. Locally invasive but non-metastasizing basal cell carcinomas may also arise in a characteristic pattern of distribution and the lesions are usually multiple. Medicinally administered arsenic often produces dose-related skin lesions. For arsenic found in drinking water, however, there is more controversy regarding the doses and exposure times necessary for cutaneous toxicity (Shannon and Strayer, 1989). Nutritional status has been implicated in the development of arsenic-related skin cancers (Hsueh et al., 1995).

Analysis of mutagenic responses of mononuclear cells (MNC) derived from patients with arsenical skin cancers revealed that at low concentration of As_2O_3 (20 μg/L arsenic) added to PHA-stimulated MNC, an amplification of the uptake of [^{3}H]-thymidine was noticed in patients with arsenical skin cancer, but not in cancers unrelated to arsenic. This observation suggests that arsenical carcinomas are hyper-reactive to arsenic's specific etiology where it seems to play a role as a co-stimulant of PHA (Yu et al., 1992). Two-dimensional polyacrylamide gel electrophoresis analysis of arsenic skin cancers—Bowen's disease, basal cell carcinoma, and squamous cell carcinoma—revealed progressive alterations in cytokeratin expression which included K16 and K6 in Bowen's disease; K16, K6 and K17 in squamous cell carcinoma and basal cell carcinoma (Yu et al., 1993).

Effects of arsenic on cytoskeleton in cultured human keratinocytes are apparently related to increased TGF alpha level resulting in a spindled morphology with long, bipolar filamentous processes and enhanced locomotion (Ju et al., 1993). Enhanced cytokine expression in arsenic-exposed keratinocytes extends also to an elevated secretion of granulocyte macrophage-colony stimulating factor, transforming growth factor-alpha and the proinflammatory cytokine tumor necrosis factor-alpha with a corresponding increase in the mRNA levels (Germolec et al., 1996). The alteration in morphology of Langerhans cells (LCs) from patients with chronic arsenicism and a significant reduction in the numbers of LCs in the Bowen's lesions as compared to the peri-lesional skin specimens suggests an arsenic-related systemic dysfunction of Langerhans cells in Bowen's disease (Wang et al., 1991). Since LCs are responsible for the immunologic presentation of tumor-associated antigens and play a role in the elimination of neoplastic clones, alteration of LC in arsenic exposure may contribute to the development of skin cancers in some patients.

Miscellaneous Tissues and Organs

Relatively little has been reported on specific organ diseases and morphological abnormalities in different tissues of animals or humans exposed to arsenic. Liver involvement in humans exposed to arsenic via ingestion pathway included hepatic angiosarcoma (Roat et al., 1982). In Wistar rats exposed to sodium arsenite (5.6 mg/kg), observed changes in the liver included hydropic degeneration, total loss of glycogen, necrosis in centrilobular zones, an increase in lipid vacuoles around the periportal area and autophagosomes containing organelle-like material, dilatation of the bile canaliculi and an increase of the number of myelin-like structures (Albores et al., 1996). In muscle cells, arsenic exposure can activate the amino acid transporter system without the involvement of stress signal transduction pathways (McDowell, Eyers, and Hundal, 1998). In porcine aortic endothelial cells in culture, induction of the inflammatory or heat-shock responses caused subtle and distinct rearrangement of the actin cytoskeleton (DeMeester et al., 1998b). Cytochalasin D alone induced a dose-dependent cytoskeletal collapse indistinguishable from that caused by the acute phase-heat shock sequence that caused cell death by apoptosis. The effect of lower doses of cytochalasin D could be potentiated by subsequent induction of the heat-shock response. It has been suggested that the cytoskeletal rearrangement is likely a critical event in the pathway to apoptosis (DeMeester et al., 1998b).

Genotoxic Effects

While arsenic is highly carcinogenic, it apparently lacks direct mutagenic activity and likely acts as a tumor promoter. Genotoxic activity of arsenic salts (sodium arsenite and sodium arsenate) has been implicated in the efficient production of methotrexate-resistant 3T6 cells, which were shown to have amplified copies of the dihydrofolate reductase gene (Lee et al., 1988). The ability of arsenic to induce gene amplification has been related to its carcinogenic effects in humans since amplification of oncogenes is observed in many human tumors. However, point mutations of codons 12, 13, and 61 in H-, K-, and N-ras oncogenes as well as in the exons 5 through 9 of the p53 tumor suppressor gene have been identified by PCR-SSCP analysis in 26 skin biopsy tissues from 16 arsenic-related Bowen's disease patients (Hsieh et al., 1994). Interestingly, analysis of biopsis taken prior to and after irradiation of skin cancer areas in Bowen's disease patients with UVB (750 mJ/cm^2) revealed that the number of p53 and Ki-67 positive cells after the UVB treatment were significantly fewer than those of non-UVB-treated specimens, suggesting that UVB irradiation has an inhibitory effect in Bowen's disease (Chai et al., 1997).

Analysis of the direct thiol-binding activity of trivalent arsenic and the resulting suppression of activity of critical enzymes associated with DNA synthesis and repair showed that direct enzyme inhibition was not a common toxic effect of As and that only a few sensitive enzymes were responsible for arsenic-induced cellular toxicity (Hu, Su, and Snow, 1998). This observation suggests that the arsenic-induced genotoxic ability and inhibition of DNA repair is probably not the result of direct enzyme inhibition, but may be an indirect effect caused by As-induced changes in cellular redox levels or alterations in signal transduction pathways and consequent changes in gene expression (Hu, Su, and Snow, 1998).

Effects on Chromosomes and Mitotic Apparatus

Chromosome Aberrations

The chromosome-breaking activity in cultured leukocytes and human skin fibroblasts is significantly higher for the compounds with trivalent than with pentavalent arsenic (Nakamuro and Sayato, 1981). Trivalent arsenic compounds inhibit the colony-forming capacity more strongly than pentavalent arsenic; however, DNA, RNA and protein syntheses are inhibited by both trivalent and pentavalent arsenic. Sister-chromatid exchange (SCE) is significantly increased in patients treated with arsenic compounds (Burgdorf, Kurvink, and Cervenka, 1977) and among individuals chronically exposed to arsenic in drinking water (130 μg/L) (Lerda, 1994). At lower concentrations of arsenic in drinking water (50 μg/L) no SCE were noted (Vig et al., 1984).

The frequency and nature of chromosomal aberrations in Chinese hamster cell lines exposed to 100-450 μg/L were assessed on the X chromosome using Giemsa and FISH techniques (Radha and Natarajan, 1998). Significant differences were noted among different cell lines studied. However, at all doses of sodium arsenite an increased number of abnormal cells with pulverised chromosomes, decondensed chromatin, isochromosomes and hyperploidy were observed in CHO9 and EM-C11 cell lines.

Evaluation of arsenic-induced genotoxicity *in vivo* is frequently made by performing the micronucleus (MN) assay on exfoliated bladder cells. These tests generally show that chromosome breakage is the major cause of MN formation when employing fluorescent *in situ* hybridization with a centromeric probe to identify the presence (MN+) or absence (MN-) of whole chromosomes within micronuclei (Moore et al., 1997). In populations exposed to elevated levels of arsenic in drinking water (600 μg/L) the prevalence of MN was found to be higher. The time factor in exposure to the same overall load of arsenic appears to play a critical role as shown in human fibroblasts treated with 90-750 μg/L arsenite for 24 h and 525-6000 μg/L for 4 h (Yih and Lee, 1999). With the same arsenite accumulation, different kinetics of cell killing and different types of MN generation were observed such that mainly kinetochore-positive MN were seen in cells at low dose exposure whereas mainly kinetochore-negative MN were induced by high dose exposure (Yih and Lee, 1999).

Mitotic Spindle Sensitivity to Arsenic

Recent experiments have demonstrated that disruption of specific cytoskeletal elements occurs in cells exposed to arsenic. Treatment of human skin fibroblasts with sodium arsenite results in arrest of cells in the G2 phase, interference with mitotic division, inhibition of spindle assembly, and induction of chromosome endoreduplication (Huang et al., 1995). Exposure of human skin fibroblasts to 375 μg/L sodium arsenite resulted in a significant delay of cell cycle progression, accumulation of mitotic cells, prolongation of mitosis, out-of-phase centromere separation and alterations of chromosome segregation, such as the appearance of c-metaphase, ball-metaphase, and lagged chromosomes (Yih, Ho, and Lee, 1997). This was accompanied by accumulation of cyclin B1 and hyperphosphorylation of cdc2 and Mos proteins. In spite of chromosomal loss in arsenite-exposed clones, they were found to have a prolonged life span (86±18 population doublings) as compared to those derived from the untreated cultures (44±11 population doublings). These observations remain to be confirmed. In lymphocyte cultures treated with 0.75 to 7.5 μg/L of As(III), a dose-dependent increase in hyperploidy was reported (Ramirez et al., 1997). Immunological analysis of the spindle apparatus with antibody against anti-beta-tubulin indicated that disruption of spindle formation was related to interaction of arsenic with the microtubules. Tubulin depolymerization and inhibition of microtubule assembly were observed also in experiments with purified tubulin.

Treatment of human cell lines HeLa S3 and KB with arsenite (375 and 750 μg/L for 24 h) arrested 35% of the total cell population at the mitotic stage, with cells displaying abnormal mitotic figures in arsenite-arrested mitotic cells, including deranged chromosome congregation, elongated polar distance of mitotic spindle, and enhanced microtubule immunofluorescence (Huang and Lee, 1998). In both cases, arsenite was found to interfere with the microtubule assembly.

Organic Versus Inorganic Arsenic

A lot remains to be learned about the role of methylation in arsenic toxicity and carcinogenesis. In most mammalian species, ingested inorganic arsenic is almost quantitatively reduced to the trivalent state in plasma and subsequently converted to the trivalent and pentavalent methylated metabolites in the liver (Thomas, Styblo, and Lin, 2001). Normal urinary excretion profile shows that dimethylarsinic acid (DMA(V)) is often the main urinary metabolite (60-80% of total arsenic concentration in urine) with monomethylarsonic acid (MMA(V)) and inorganic arsenic each accounting for 10-20% of the urinary arsenic (Vahter, 1999; Schwerdtle et al., 2003). For a long time, in vivo biomethylation and excretion of inorganic arsenic in the form of MMA(V) and DMA(V) were thought to be a major detoxification process. Recent studies, however, show that in addition to the pentavalent metabolites, both monomethylarsonous acid (MMA(III); 2-5%) and dimethylarsinous acid (DMA(III); 4-21%) can be detected in urine of people chronically exposed to inorganic acid via drinking water in West Bengal, India (Mandal, Agra, and Suzuki, 2001). The biological effects of trivalent methylated metabolites include enzyme inhibition, cell toxicity and genotoxicity (Thomas, Styblo, and Lin, 2001). There are thus three possible outcomes of methylation as a biotransformative process that may alter arsenic toxicity or carcinogenicity: (i) methylation is protective by inducing detoxification and excretion of arsenic from the body thereby lowering tissue-specific dose and toxic effects of inorganic arsenic; (ii) methylation forms more toxic metabolites of arsenic that are more reactive with tissues and have carcinogenic properties; (iii) methylation of arsenic creates competition with DNA methylation by competing for methyl donation from s-adenosyl methionine (SAM) thereby leading to DNA hypomethylation (Moore et al., 2003). Although much progress has been made in understanding the mechanisms of arsenic toxicity and carcinogenicity, a scientific consensus has not yet been reached.

Significant differences in the effects of organic and inorganic arsenic compounds on cellular metabolism, morphology and growth have been reported. With respect to their effects on cell division in cultured Chinese hamster V79 cells, inorganic arsenic compounds, which are contaminants found in drinking water, inhibited cell growth at very low exposure doses. In contrast, methylated derivatives of inorganic arsenic such as monomethylarsonic acid (MMA), dimethylarsinic acid (DMA), and trimethylarsineoxide (TMAO), were less cytotoxic (Eguchi, Kuroda, and Endo, 1997), and their cytotoxicity was further decreased with an increase in the number of methyl groups in the compounds. Arsenobetaine (AsBe) and arsenocholine (AsC), which have been found in some marine products, did not show any cytotoxicity. However, an important difference was found in the ability of organic and inorganic arsenic compounds to interfere with mitosis and induce ploidy aberrations. Both DMA and TMAO induced tetraploidy in exposed cells, whereas arsenite, arsenate, AsBe and AsC did not induce mitotic arrest or tetraploids. This ability of organic arsenic to produce aneuploidy suggests that MMA, DMA and TMAO exert some effect on cell division during the metaphase that may provide insight into the carcinogenic mechanism of arsenic (Eguchi, Kuroda, and Endo, 1997).

The higher mitotic toxicity of organic arsenic in some studies can be attributed to the more significant disruptive effects of this form of arsenic compared to inorganic compounds on the microtubular organization of the cell. In Chinese hamster V79 cells, DMA produced much more significant changes in cytoskeletal organization, mitotic arrest and induction of multinucleated cells than the inorganic arsenic (Ochi, Nakajima, and Fukumori, 1998). The mitotic arrest caused by DMA was accompanied by disruption of the microtubule network. The toxicity appears to be selectively targeted at the microtubular network since neither arsenite nor DMA causes disorganization of actin stress fibers even at concentrations that produce a marked retardation of cell growth. There are differences in the effect of organic and inorganic arsenic on DNA synthesis; for instance, cells exposed to arsenite for 6 h showed marked inhibition of DNA synthesis, whereas inhibition by DMA was not observed. It appears that organic arsenic (DMA) primarily causes growth retardation and morphological changes via disruption of the microtubule network, whereas arsenite-induced retardation of cell growth and inhibition of DNA synthesis does not occur via disruption of the cytoskeleton (Ochi, Nakajima, and Fukumori, 1998).

The differences in cellular toxicity of organic and inorganic arsenic compounds are certainly not universal. Apparently, the toxicity varies with the specific compound and the target cell. For instance, in myeloid leukemia cell lines, both inorganic and organic acid compounds have been shown to inhibit cell growth, induce apoptosis, and downregulate bcl-2 protein in all cell lines tested (Wang et al., 1998). Arsenic trioxide synergistically with interferon-alpha induced cell cycle arrest and apoptosis in adult T-cell leukemia/lymphoma cells (Bazarbachi et al., 1999). Efficiency of inhibition of cell growth and DNA synthesis among different organic arsenic compounds including arsenic-containing chemical warfare agents was suggested to depend upon solubility of the compound rather than on its arsenic content (Henriksson et al., 1996).

Generation of Oxidative Stress and Heat Shock Response by Arsenic

While separation of oxidative stress and heat shock response is rather artificial since those two processes are intimately interrelated in many instances, findings primarily focused on oxidative stress phenomena will first be discussed followed by those on the heat shock response in arsenic-exposed cells.

Arsenic-Induced Oxidative Stress

The most common form of arsenic in the environment is arsenate. Because of the similarity in oxygen coordination and overall geometry, arsenate often acts as a phosphate mimic and has been shown to selectively inhibit mitochondrial oxidative phosphorylation. By contrast, arsenite enters the cell through simple diffusion and its toxicity is believed to be related to thiol reaction. By targeting sulfhydryl goups within a critical protein molecule, it can denature the protein leading to autophagic degradation and an increase in reactive oxygen species (ROS). Any iron released during protein denaturation can participate in a Fenton reaction thereby generating hydroxyl radicals, which are known to damage the DNA (Lee and Ho, 1994). Arsenite thus elicits its toxicity indirectly by binding and inhibiting a number of enzymes, which are necessary for removal of ROS including superoxide dismutase (removal of superoxide) and catalase (removal of hydrogen peroxide, H_2O_2). Due to cellular respiration, continuous production of ROS as well as H_2O_2, superoxide anion and peroxynitrite can occur in the mitochondria (Parker and Murphy, 1995). Intracellular ROS, including H_2O_2, hydroxyl radical, singlet oxygen and superoxide anion have all been shown to directly or indirectly damage critical molecules such as proteins, lipids, and DNA. The most likely cellular damage produced by mitochondrial ROS is believed to be due to oxidative stress, which creates an imbalance between the concentration of ROS in the cell and the ability of the cell to scavenge or detoxify it.

These oxygen radicals can cause apoptosis of human umbilical vein endothelial cells (HUVECs) after a typical stress such as heat shock that can be induced by either hyperthermia or sodium arsenite (Jacob et al., 1997). Iron-catalyzed formation of ROS appears to be critical for inducing cell death, because both the chelation of iron and the enzymatic detoxification of ROS attenuate apoptosis. Along with destructive effects, sodium arsenite apparently can induce an enzyme (heme oxygenase-1) that catalyzes heme degradation in chicken hepatoma cells. This enzyme protects cells against oxidative stress-related injury (Elbirt et al., 1998). The signal cascade triggered by sodium arsenite and heat shock, but not heme, leads to the induced activity of the MAP kinases extracellular-regulated kinase (ERK), c-Jun N-terminal kinase (JNK), and p38. Eventually, this pathway induces heme oxygenase-1 at the genomic levels with the involvement of at least one AP-1 element (Elbirt et al., 1998).

The rate of apoptosis in human polymorphonuclear neutrophil leukocytes (PMN) assessed qualitatively by morphology or gel electrophoresis of DNA and quantitatively by CD16 receptor expression and propidium iodide DNA staining often shows an increase on incubation with arsenite (Watson et al., 1996). The arsenite-induced PMN apoptosis correlates with an increase in intracellular respiratory burst, an oxygen-dependent mechanism, and can be prevented using selective antioxidants such as glutathione (GSH), N-acetylcysteine (NAC) and taurine. However, catalase and dimethylsulfoxide (DMSO) fail to provide protection.

Consistent with these observations, arsenite-induced apoptosis can be abrogated by administration of oxygen radical scavengers, suggesting that heat shock-induced programmed cell death in LPS-primed endothelial cells can be arrested, at least in its early stages, by agents that augment or stabilize the reducing potential of the cell (Abello, Fidler, and Buchman, 1994). Comparison of arsenite- and heat-induced thermotolerance related to protein denaturation and protein aggregation in cells reveal that the two agents produce comparable levels of resistance to subsequent heat, arsenite-, ethanol- and diamide-induced toxicity although levels of the major HSPs induced by heat [hsp27, hsp70(p73), hsp70(p72), and hsp90] and arsenite may be different (Kampinga et al., 1995). Interestingly, pretreatment of both V79 and As/R28A cells (an arsenite-resistant Chinese hamster V79 cell variant) with low doses of sodium arsenite is accompanied by both an increased thermotolerance and arsenic tolerance (Wang, Hou, and Rossman, 1994). However, thermal pretreatment does not increase arsenic tolerance, suggesting that induction of thermotolerance and arsenic tolerance by arsenite pretreatment apparently occurs by different mechanisms (Wang, Hou, and Rossman, 1994). Protective effects of HSPs against non-thermal stresses are well known. An example is the induction of HSP 70 which protects cells from radiation-induced apoptosis by down-regulating p53 and *bax* expression (Gordon et al., 1997).

The mechanism of MN formation in arsenic-exposed cells has been traced to the generation of reactive oxygen species during the metabolism of arsenic compounds (see above). The MN formation has been induced in a dose-dependent manner by a 4-h treatment with arsenite. In particular, the arsenite-induced nitric oxide (NO) increase suggests the involvement of peroxynitrite in arsenite-induced MN which can be suppressed by NO synthase inhibitors and calcium chelators (Gurr et al., 1998). It would appear that the disturbance of NO production may cause cardiac and vascular disorders, and the peroxynitrite-mediated DNA damage may cause genetic instability and, hence, cancers in arsenic-exposed humans. In laboratory animals, administration of radical scavenging substances (mustard oil and garlic extract) allegedly prevented chromosomal aberrations induced by sodium arsenite (Choudhury, Das, and Sharma, 1997; Choudhury et al., 1997).

The important role of the oxidative stress-related effects in the arsenite-induced heat shock response is revealed in experiments that studied the effects of reducing agents, dithiotreitol and mercaptoethanol, on the expression of HSPs. The expression of small HSPs (HSP27, alpha B crystallin) and larger HSP70 protein family induced by sodium arsenite in C6 rat glioma cells occurs in two phases. One phase of the heat shock response can be enhanced by low concentration of mildly reducing dithiothreitol (2-mercaptoethanol). The other phase of the heat shock response leads to activation of the heat shock factor at the genomic level and is very sensitive to suppression by a reducing agent such as dithiothreitol which is present during the stress period (Kato, Ito, and Okamoto, 1997).

Arsenic-Induced Heat Shock Response

Like heat shock, arsenic is capable of inducing HSPs of various sizes from the smaller proteins including alphaB-crystallin, HSP25, HSP27, HSP30, etc (Huot et al., 1995; Haynes, Duncan, and Piatigorsky, 1996; Ito et al., 1997; Ohan et al., 1998; Zavialov et al., 1998; Lambert et al., 1999) to the larger HSP species such as HSP105 (Yasuda et al., 1995). New heat shock proteins are being continually identified in cells exposed to a variety of stress agents. A new p28, small stress protein with homology to glutathione-S-transferase, is produced in response to cytotoxic agents and plays a role in cellular redox homeostasis (Kodym, Calkins, and Story, 1999). Another recently discovered, very abundant HSP15 binds directly to DNA and RNA (Korber et al., 1999).

The Hsp27 is involved in the activation process of protein kinase B in the signal transduction pathway of various forms of stress (Konishi et al., 1997). Of the many HSPs synthesized in cells exposed to environmental and other type of stresses, HSP72 appears to be the most stress-inducible one (Wijeweera et al., 1995). It is induced to a different extent by different agents (heat, sodium arsenite, UV, γ-irradiation) in normal and transformed cell lines but no HSP72 induction is detected in any of the cell lines after gamma-ray irradiation (Muramatsu et al., 1995). HSP72 can be induced not only by inorganic arsenite or heat shock, but also by dimethylarsinic acid (DMA), an important metabolite of arsenics in mammals, however, the nuclear distribution of HSP72 when induced by DMA is different than when it is induced by inorganic arsenic or heat shock (Kato et al., 1997) as well as by other stressors (Sacco et al., 1997).

It has been suggested that HSP expression is generated by abnormally folded, non-native proteins produced by stress (Gabai et al., 1997) capable of triggering oxidation of non-protein thiols, particularly glutathione, and leading to heat shock transcription factor (HSF) activation (Zou et al., 1998). Activation of the HSF is produced by hyperphosphorylation subsequent to stress and is mediated via mitogen-activated protein kinases (MAPK) family members such as c-Jun NH_2-terminal kinase (JNK). JNK signal transduction pathway, also referred to as stress-activated protein kinase (SAPK) pathway, can be efficiently activated by oxidative stress (for a review see Feder and Hofmann, 1999).

Activation of the HSF leads to an increase in the HSP70 family gene expression (Gomer et al., 1996; Lee and Corry, 1998). On the other hand, over-expression of HSP70 inhibits the activation of its transcriptional factor HSF-1 (Ding, Tsokos, and Kiang, 1998). HSF-1, in turn, inhibits the expression of c-fos, an immediate early gene that controls responses to extracellular stimuli for growth and differentiation (Chen et al., 1997). Inducibility of HSPs in cells exposed to thermal or chemical stress varies with the state of differentiation. Sodium arsenite, like tumor necrosis factor, enhances the expression of a heat-inducible transformation-associated cell surface antigen expressed on the activated H-ras oncogene-transformed rat fibrosarcoma W31, which is recognized by antibodies, to HSP70 (Tamura et al., 1993). In neuronal cells, for instance, this response is attenuated relative to the level of HSP70 induction in undifferentiated cells, which may explain the sensitivity of neuronal cells to pathophysiological stressors (Hatayama, Takahashi, and Yamagishi, 1997).

The very fact that the heat shock protein response can be induced not only by heat shock but also by a variety of different chemicals suggests that a common denominator of these different inducers may exist. Chemicals and conditions that damage proteins, promote protein misfolding, or inhibit protein processing appear to trigger the onset of protective homeostatic mechanisms resulting in "stress responses" in mammalian cells. It has been suggested that all inducers may generate abnormally folded, i.e. non-native, proteins, which may trigger the heat shock response (Brostrom and Brostrom, 1998). One common mechanism leading to such a modification of proteins is seen in the fact that all chemical inducers similar to thermal shock cause drastic oxidation of glutathione under conditions which they also induce shock promoter element (HSE), DNA-binding activity and cause trimerization of HSF-1 (Zou et al., 1998). Similarities in metabolic responses of cells exposed to arsenic and hyperthermia extend from the peripheral stages of the signal transduction pathway to transcriptional regulation of induction of heat shock proteins.

Transcriptional Regulation by Arsenic Stress

The complexity of interaction of HSPs with transcription regulatory factors is being actively investigated since it offers an insight into important pathways controlling cellular response to the environment, adaptation, growth stimulation and inhibition, differentiation and cell death. Activation of the HSF is linked to the synthesis of HSP70. HSF activates expression of its target genes in response to hyperthermia, chemical or physiological stress. Under non-challenged conditions, HSF is maintained in an inactive non-DNA-binding form and acquires the ability to bind specifically to the heat shock promoter element (HSE) in response to elevated temperatures or other conditions such as arsenic exposure (Mosser, Duchaine, and Massie, 1993).

A key step in the activation process involves the formation of HSF homotrimers, leading to high-affinity DNA binding. Trimerization and DNA binding of purified HSF can be directly and reversibly induced by heat shock temperatures in the physiological range and by an oxidant such as hydrogen peroxide (Zhong, Orosz, and Wu, 1998). Other chemical inducers of a heat shock response, including salicylate, dinitrophenol, ethanol, and arsenite, do not cause HSF trimerization in vitro, indicating that these inducers act by indirect mechanisms (Zhong, Orosz, and Wu, 1998). However, different HSF (HSF-1 and HSF-3) have different thresholds to the stress stimuli (Tanabe et al., 1997).

Upregulation of HSF-1 occurs through phosphorylation of HSF-1 that can be induced by various stressors known to increase the synthesis of HSPs such as heat shock (43°C for 1 h), hypoxia (5 h exposure to 0.02% oxygen), 8% ethanol (1 h at 37°C) or sodium arsenite (1 h at 37°C) (Mivechi et al., 1994; Kim et al., 1998). Subsequent binding of HSF to heat shock response element, that causes transactivation of heat shock mRNAs, is controlled by various factors including stress-inducible and developmentally regulated events (Gordon et al., 1997). While upregulation of HSF by heat and sodium arsenite appears to be similarly effective in reducing the effectiveness of the MDR-1 gene expression and enhancing sensitivity of resistant cells to drugs (Kim et al., 1998), experiments with calcium ionophore A23187 suggest that the pathways of HSF phosphorylation induced by heat and arsenic are different, implying distinct mechanisms of transcriptional control by heat and arsenic (Elia et al., 1996).

One of the early gene expression events in response to various stressors including heat shock, arsenite and other agents is the induction of the early transcription regulator factor—the proto-oncogene c-fos that is important in cell growth, differentiation and development. Several promoter elements have been identified as mediators for the induction of c-fos through interaction with the heat shock element by serum, tumor promoting agents, epidermal growth factor and other stimuli. Arsenite-related induction of c-fos is also mediated by the serum response element (van Delft et al., 1993). Human and rodent c-fos promoters have recently been shown to contain HSEs which are highly responsive to heat, arsenite and cadmium (Ishikawa et al., 1999).

It is significant that the immediate early genes c-fos, c-jun and egr-1 usually activated by growth factors, cytokines, differentiation signals and DNA damaging agents can be activated by arsenic-induced stress via the JNK (SAPK) and p38 signal transduction pathways (Kawasaki et al., 1996; Lim, Jain, and Cao, 1998). The general nature of activation of this signal cascade by arsenic can be seen in its ability to stimulate binding to DNA of the insulin upstream factor 1 (IUF1), a transcription factor present in pancreatic beta-cells, and IUF1-dependent gene transcription (Macfarlane et al., 1997).

Trivalent arsenic As(III), but not As(V), is a potent stimulator of AP-1 transcriptional activity and an efficient inducer of c-fos and c-jun gene expression (Cavigelli et al., 1996). Induction of c-jun and c-fos transcription by As(III) is related to activation of JNKs and p38/MAPK2, which phosphorylate transcription factors that activate these immediate early genes. As(V), on the other hand, has a negligible effect on JNK or p38/ MAPK2 activity. The primary mechanism by which As(III) stimulates JNK activity apparently involves the inhibition of a constitutive dual-specificity JNK phosphatase. This phosphatase activity appears to be responsible for maintaining low basal JNK activity in non-stimulated cells and its inhibition may lead to tumor promotion through induction of proto-oncogenes such as c-jun and c-fos and stimulation of AP-1 activity (Cavigelli, et al., 1996).

Transcriptional regulation in HeLa cells by either heat shock or sodium arsenite involves enhancement of phosphorylation of the C-terminal domain of the largest subunit of eukaryotic RNA polymerase II that serves as a signal for initiation of transcription (Dubois et al., 1994). Arsenic, along with heat shock and ethanol, is able to transcriptionally activate via HSF-1 the expression of the beta-amyloid precursor protein gene involved in the pathogenesis of Alzheimer's disease (Dewji and Do, 1996). Sodium arsenite, like heat shock, is able to modulate glucocorticoid receptor (GR) status, reproducing the effect of hormone binding to the receptor, indicating that cellular stress can cause at least a partial induction of hormone-independent GR-mediated gene expression (Shen et al., 1993; Sanchez et al., 1994). Furthermore, sodium arsenite like a heat shock, enhances GR-mediated gene expression (Hu, Guan, and Sanchez, 1996). However, in spite of similarities in transcription regulation by heat shock and sodium arsenate, they display distinct differences, at least in mouse cells, at the level of splicing of HSP47 mRNA such that temperature shock can induce alternative mRNA splicing, whereas sodium arsenite and amino acid analogs cannot (Takechi et al., 1994).

At the level of transcription regulation, arsenite, similar to other reactive oxygen species generators, induces nitric oxide production, potently induces poly(ADP-ribosylation), NAD depletion, DNA strand breaks, and micronuclei in CHO-K1 cells (Lynn et al., 1998). While NF-κB is activated by an inducer of the acute phase response (lipopolysaccharide), pretreatement with sodium arsenite induces IκB-alpha (the cytoplasmic inhibitor of NF-κB) and reduces subsequent apoptosis provoked by lipopolysacharide (LPS) (DeMeester et al., 1997). Inhibition of NF-κB activity by pyrrolidine dithiocarbamate (PDTC) has the same effect as exposure of cells to sodium arsenite that results in apoptosis. It appears, therefore, that inhibition of the transcription factor NF-κB, an important regulator of the cellular response to diverse stresses, may act to actually produce heat shock response in cells (DeMeester et al., 1998a). At the same time, the treatment with arsenite, heat shock or addition of the nuclear factor κB (NF-κB) inhibitor, pyrrolidine dithiocarbamate (PDTC), significantly attenuates cytokine-induced NOS2 mRNA and NO production in rat hepatocytes (de Vera et al., 1996a, b).

Interaction of the Heat Shock Response With Other Metabolic Systems

Clearly, there are mechanisms operating at levels other than transcriptional regulation in cells responding to various stressors. The relationship of the HSP27 induction to other metabolic pathways is demonstrated in coupling of the response to stress with the metabolic activity of the arachidonic acid cascade—the activation of protein kinase C (PKC) inhibits the induction of hsp27 through the suppression of arachidonic acid release (Suzuki et al., 1996). On the other hand, HSP27 induction by arsenite or thermal stress is linked to activation of protein kinase B (Konishi et al., 1997) that is mediated by phosphoinositide 3-kinase rather than MAPK activated protein kinase (Shaw, Cohen, and Alessi, 1998).

Early in the pathway of the transmission of stress signals in the cytoplasm, the AMP-activated protein kinase is believed to be the central component of a protein kinase cascade that phosphorylates and inactivates key regulatory enzymes of several biosynthetic pathways which act protectively, switching off biosynthetic pathways when the cell is subjected to stress that causes ATP depletion, with the key signal being a rise in AMP level (Corton, Gillespie, and Hardie, 1994). It appears that regulatory pathways controlling cellular response to different type of stress may converge on a link between intracellular Ca^{2+} homeostasis and phosphorylation of key intracellular proteins (e.g., eukaryotic initiation factor-2alpha) as seen in arsenite-challenged cells (Brostrom et al., 1996; Senisterra et al., 1997). In fact, calcium ionophore A23187 inhibits HSP70 expression in human cells induced by either heat or sodium arsenite (Elia et al., 1996). Phosphorylation of HSP27 in human keratinocytes exposed to sodium arsenite is related to intracellular Ca^{2+} homeostasis and may be mechanistically linked to control of keratinocyte growth and differentiation as well as responses of keratinocytes to extracellular stresses (Shi and Isseroff, 1996).

Stressor-Specific Features of the Heat Shock Response

While there are different chemical agents capable of eliciting a heat shock response through the induction of various HSPs, the kinetics of induction may vary depending on the chemical. This is the case, for instance, with a member of the HSP70 family, HSP72, that appears in rat embryos exposed to some chemical teratogens, including sodium arsenite, but not when other compounds such as N-acetoxy-2-acetylaminofluorene, cadmium chloride and cyclophosphamide are used (Mirkes, Doggett, and Cornel, 1994). Specificity of the cellular response to different stressors can be manifested in the induction of some HSPs that are not produced by other stressors. For instance, comparison of induction of HSPs in rat hepatoma cells by heat shock, arsenite, cadmium, dinitrophenol and ethanol reveals that some agents fail to stimulate specific HSPs to a degree that is comparable to the induction of these HSPs by heat shock (Wiegant et al., 1994). The HSP100 is hardly induced by sodium arsenite, whereas treatment with DNP only slightly induces HSP68 and HSP84, and no detectable induction of HSPs is observed in these cells after treatment with ethanol (Wiegant et al., 1994). Similarly, in *Xenopus laevis,* sodium arsenite stress induces HSP90 expression while cadmium fails to do so (Ali et al., 1996). Despite stress-specific response mechanism controlling induction of specific HSPs by various stressors, there are commonalities in the regulation of the expression of at least some HSPs (Wiegant et al., 1996).

In Reuber H35 hepatoma cells exposed to increasing doses of heat, arsenite or cadmium capable of activating the heat shock transcription factor (HSF), the increases of mRNA levels and rates of synthesis of HSP60, HSP68, and HSP84 showed stressor-specific patterns (Ovelgonne, Bitorina, and Van Wijk, 1995). In contrast to heat and arsenic that induced HSPs, cadmium ions failed to induce either HSP60 or its mRNA, although under these conditions maximal activation of HSF was observed (Ovelgonne, Bitorina, and Van Wijk, 1995). This, however, may be related to the specifics of the dose-response pattern of the hepatoma cells, since in CHO cells, for instance, synthesis of the inducible form of HSP70 could only be achieved at higher doses of arsenic and cadmium ions (Liu, Corry, and Lee, 1995). When initial stress is produced by different agents such as hyperthermia, sodium arsenite or cadmium ions capable of inducing a stressor-specific pattern of expression of HSPs, the subsequent stress response elicited by lower doses of these stressors does not depend on the preconditioning dose, but on the subsequent low level stressor (Wiegant, Spieker, and Van Wijk, 1998). These observations are important for selecting and using HSPs as biomarkers of environmental stress.

Protective Role of the Heat Shock Response

The heat shock response produced either by hyperthemia or chemical agents such as arsenic is a highly conserved stress response known to alter patterns of gene expression and appears to be able to modulate the ability of cells to subsequently respond to other inducing agents such as cytokines (Abello and Buchman, 1994; Wong et al., 1995). Ability of sodium arsenite to induce HSP expression is part of the stress response and constitutes a defensive mechanism of cells against similar or different type of stress. Regulatory mechanisms involved in heat shock response to different levels of stress can provide protection to cells exposed to subsequent stress challenges, to the extent of preventing a more severe damage or even death. A case in point is the ability of a heat shock or sodium arsenite to induce the expression of HSP72 in human endothelial cells (EC) that prevents the development of subsequent necrosis by activated human polymorphonuclear neutrophil leukocytes (PMNs) (Wang et al., 1995). Lower concentrations of sodium arsenite did not cause EC necrosis but significantly attenuated subsequent EC necrosis induced by activated PMNs. Higher doses of sodium arsenite led to dose-dependent apoptosis with the highest dose of sodium arsenite tested causing a significant increase in EC necrosis (Wang et al., 1995).

In cultured sheep pulmonary artery endothelial cells (SPAEC), apoptosis can be induced by lipopolysaccharide (LPS) and the heat shock response produced either by elevated temperature or exposure to sodium arsenite significantly counteracts the apoptotic changes (Wong et al., 1996). Both treatments resulted in induced expression of HSP70. Transient over-expression of HSP70 produced by direct gene transfer protected the cells from apoptosis, suggesting a direct role for HSP70 in the attenuation of LPS-mediated apoptosis (Wong et al., 1996).

Such a protective capability against developing a heat shock response upon exposure to arsenic has also been observed in human skin (Muramatsu et al., 1992). At low concentrations, sodium arsenite alone, much like heat shock alone, still fails to produce apoptosis in endothelial cells. However, when these agents are combined with oxidative stress, apoptosis takes place (Wang et al., 1997). Induction of heat shock response by sodium arsenite prevented endotoxin (LPS)-challenged rats from hypotonic shock mediated by inducible nitric oxide synthase (iNOS) (Hauser et al., 1996). Similarly, pre-treatment with sodium arsenite prevented cytokine-inducible nitric oxide synthase expression in rat hepatocytes (de Vera et al., 1996a, b). The synthesis of Sendai virus in African green monkey kidney cells was blocked by various stressors such as sodium arsenite, cadmium and heat shock, all of which were capable of inducing HSP70. The latter serves as a mediator of the antiviral effect in these cells, resembling induction of HSP70 expression in response to prostaglandin A1 administration (Amici et al., 1994).

Sodium arsenite itself effectively induces the expression of HSP72 in different organs (lung, liver, kidney, and duodenum) and can improve the survival of female ND4 mice from endotoxin (Lappas, Karl, and Hotchkiss, 1994). Rats are also protected by sodium arsenite against sepsis through the induction of HSP72 expression in the lung (Ribeiro et al., 1994). On the other hand, inhibition of HSP70 expression makes cells much more vulnerable to arsenite (Dix, Garges, and Hong, 1998).

In a mechanism common to that operating in thermal stress, arsenite strongly enhances serine phosphorylation of a small heat shock protein (HSP27) in human diploid FS-4 fibroblasts (Vietor and Vilcek, 1994; Huot et al., 1995; Ito et al., 1997). Tyrosine phosphorylation of signal transducing proteins is a feature of the arsenite-induced stress shared with other stressors. It involves the mitogen-activated protein kinase (MAPK)/extracellular-signal regulated kinase (p38MAPK) cascade (Rouse et al., 1994; Masuya et al., 1998). However, distinctions exist between the specific MAP kinase pathways followed during stresses of different nature. Arsenite, like phorbol esters and the cytokines tumor necrosis factor-alpha and interleukin-1-beta can cause an increase in phosphorylation of eukaryotic initiation factor eIF4E by the specific p38 MAP kinase. However, other stresses such as heat shock, sorbitol, and H_2O_2, which also stimulate p38 MAP kinase and increase Mnk1 activity, do not increase phosphorylation of eIF4E (Wang et al., 1998). Elevated levels of HSP72 inhibit a signal transduction pathway leading to programmed cell death by preventing stress-induced activation of JNK while stress-induced activation of another stress-kinase, p38 (HOG1), is also blocked when the level of HSP72 is increased (Gabai et al., 1997). Experimental data suggest that, in regulation of JNK and p38 kinases, HSP70 family of factors serve as a "sensor" of the build-up of abnormal proteins after heat shock and other stresses.

Another damage control system that operates as part of the stress response is the metallothionein induction implicated in protecting cells exposed not only to heavy metals (Bittel et al., 1998; Chu et al., 1999), but any stress, including oxidative stress (Ghoshal et al., 1998). It is important to note that the protective effects of the expression of HSPs extend beyond the cellular level. As shown in rats, HSP72 induced either by pre-exposure to heat or injection of sodium arsenite protects against a whole body stress such as heat stroke (Yang et al., 1998).

It is apparent from the review above that the intracellular effects of arsenic represent prototypical stress response of cells with the involvement of the "classical" heat shock components. The general molecular events in the signal cascade induced by arsenic now are being elucidated in fine detail. Nevertheless, very little is still known about the arsenic-specific signal transduction events as well as the genomic responses and cellular mechanisms involved in counteracting and repairing arsenic-induced injury. Information obtained in molecular studies of other stressors can be usefully applied to the analysis of arsenic effects in controlled cell systems. Hopefully, understanding of the effects of arsenic at the molecular genetic level with in vitro studies will help develop early detection of and preventive measures against arsenic exposure in humans.

CHAPTER 2
MATERIAL AND METHODS

GENERAL

Two different but well characterized *in vitro* skin models have been used in our laboratory for many years in the study of dermal toxicity of chemicals. Protocols have been developed for routinely obtaining primary human (Bernstam, Vaughan, and Bernstein, 1990) and animal (Vaughan, Gray, and Bernstein, 1986; Bernstam, Vaughan, and Bernstein, 1986) stratified, cornified cutaneous cultures, which mimic the structure and have most of the biochemical characteristics of the epidermis *in situ*. This study relies on artificial skin produced from human keratinocytes and grown on the Nylon Puropore membranes (Gelman Sciences).

Ordered stratification and extensive cornification is obtained in our lab by "lifting" the culture to the air-liquid interface. Lifted cultures of both human and animal keratinocytes develop barriers to the permeation of water as demonstrated by the values (Kp) measured for the permeation of ^3H_2O (Pu, et al., 1995). Cultured keratinocytes can be kept as a submerged monolayer by restricting the level of Ca^{2+} in the medium (Hennings et al., 1980). Although stratification is retarded under these conditions, differentiation to the lower spinous stage does occur (Brown, Ku, and Bernstein, 1987). The normal kinetics of DNA replication and mitosis in this culture were defined for the first 7 days after seeding (Ku and Bernstein, 1988). Essentially all cells undergo replication of DNA followed by mitosis within the first 3 days in culture, while only about 60% of the cells are proliferative by 6 days in culture. The remaining fraction of cells exhibit early differentiative characteristics even though the culture continues to be a monolayer.

Uniqueness of "Lifted" Keratinocyte Cultures for Toxicity Studies

The mode of growing keratinocytes as a "lifted" culture affords several unique opportunities to the investigator. Specifically, such cultures can be exposed to a toxic agent under study in two distinct ways, namely:

a. an agent under study (arsenic compounds) can be applied on the surface of the culture, thereby offering a skin model with a topical application of an agent;
b. alternatively, an agent under study can be added to the culture growth medium so as to reach the basal cells from underneath, thereby simulating this particular route of entry of toxic compounds to the skin in a human body.

The advantage of culturing keratinocytes by either process allows us to observe cellular response to arsenic resembling the diversity of *in vivo* exposures in humans. This approach may offer an advantage in elucidating the gradual process of the development of skin carcinomas stemming from transformation of basal cells, reflecting otherwise undetectable alterations occurring early in skin carcinogenesis. In contrast, submerged cultures limit administration of toxic agents under study to only one mode of application, i.e., by adding them to the growth medium.

Source of Tissue

Human keratinocytes (HK) and fibroblasts (HF) were isolated from adult human skin yielded during cosmetic female breast reductive surgery. Human skin specimens (from 20- to

60-year olds) were obtained from the University of Michigan Hospitals and Cooperative Human Tissue Network. For the purposes of this study, no selection of specific ethnic or age groups were made. The study was conducted in accordance with national and institutional quidelines for the protection of human subjects.

Preparation of Cell Cultures

The protocol used for isolation, plating and growth of HK was described in detail in Bernstam et al. (2002). Briefly, porcine crystalline trypsin (Sigma, type IX) was used for 16 hr at room temperature to separate the epidermis from the dermis. Epidermal cells were collected into minimum essential medium (MEM) by scraping the dermis with a scalpel. The keratinocytes were purified on a continuous gradient formed by centrifugation of 38% Percoll® (Pharmacia) at 30,000 g for 15 min at 4°C (Ribeiro, Mitra, and Bernstein, 1991). The lower band (density>1.087g/cc), consisting of a population of germinative cells, was collected, rinsed, pelleted, and resuspended in MEM. Viable cells were counted using trypan blue exclusion dye technique and resuspended in complete growth medium (MEM, supplemented with 10% fetal bovine serum (FBS) (ICN Biomedical), 10 µg/ml insulin, 10 µg/ml hydrocortisone, 100 units/ml penicillin, 100 µg/ml streptomycin, 0.05 µg/ml fungizone® (all from Gibco), and 10 ng/ml epidermal growth factor (EGF) (Becton Dickinson). Bovine pituitary extract (BPE) was prepared from frozen bovine pituitaries (Pel-Freeze®, France) by the method of Boyce and Ham (1983), and 60 µg/ml was added into the medium. The keratinocytes were then seeded onto plastic dishes (Corning) to obtain submerged cultures or onto Nylon Puropore® membranes (Gelman Sciences, 0.2 µm pore size, 13 mm diameter) in order to obtain cultures grown on the air-liquid interface (artificial skin). Both plating surfaces were precoated with 6 µg/cm^2 of 0.03% calf skin collagen, type 1, soluble in acid (Sigma). Plating density was 1×10^6 cells per cm^2 for submerged and lifted cultures. Cells were grown at 35°C, 5% CO_2 and 95% humidity. Medium was changed 24 hr after plating and every second day after that. In order to grow stratified, cornified cultures of cutaneous keratinocytes, the skin cells, plated on the Nylon Puropore® membranes, were first grown submerged for 5 days and then "lifted" to the air-liquid interface by placing the membranes to the glass fiber filters (Pall-Gelman Sciences), saturated with the growth medium (Figure 2.1). Cultures were incubated for an additional 10–15 days. Concentrations of EGF and BPE in the growth medium were increased 1.5 times, and 0.8 ml medium was changed daily during the period of growth at the air-liquid interface.

Human fibroblasts were isolated from the dermis of the same samples of the skin used for obtaining the keratinocytes. Dermis was cut into 4–5 mm^2 squares and incubated for 20 min with stirring at 35°C in freshly-made filter-sterilized 1:1 mixture of 0.1% collagenase (Worthington) and Dispase (Boehringer Mannheim) dissolved in Earle's Balanced Salt solution (EBSS). The resulting cell suspension was collected, centrifuged at 30g for 10 min, resuspended in MEM, supplemented with 10% FBS (ICN Biomedical), 10 µg/ml insulin, 10 µg/ml hydrocortisone, 100 units/ml penicillin, 100µg/ml streptomycin, 0.05 µg/ml Fungizone® (all from Gibco) and plated into culture flasks. The residual dermis was stirred in complete MEM for 1 hour at 37°C. Cells were collected and plated as above. This procedure was repeated two more times at room temperature. Cell numbers in suspension were not determined for lack of reliability of counts at this stage. The medium was changed 24 hrs later and every second day subsequently. When cultures were 70-80% confluent, cells were subcultivated and used for experiments 24 hrs later.

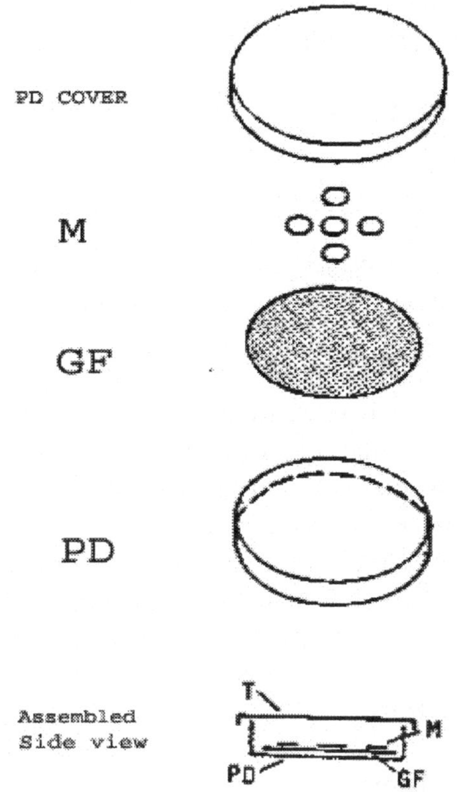

Figure 2.1 Schematic diagram illustrating method for obtaining artificial skin. Cells are seeded on membranes (M) and grown in submerged cultures for 5 days. The membranes are transferred to a Petri dish (PD) and placed on the surface of glass fiber filter pads (GF) saturated with nutrient medium where the culture on the membrane continues to differentiate at the air-liquid interface.

EXPERIMENTAL DETAILS: EXPOSURE OF CELLS TO ARSENIC

Stock solutions of As(III) and As(V) were made in Milli-Q water and the concentration of As was confirmed by graphite furnace atomic absorption spectrometry (GFAAS) or inductively coupled plasma-mass spectrometry (ICP-MS). Sodium arsenate (As(V)) and sodium metaarsenite (As(III)) used for the stock solutions were obtained from Aldrich, St. Louis, MO. Effects of buffer reagents (NaCl, EBSS, Solution A) on arsenic determination was evaluated, and the least interference was found with NaCl matrix.

Cornified, stratified cultures, grown on nylon microporous membranes were exposed for varying lengths of time to arsenic solution by topical application of 80 μL of known concentrations of As(V) or As(III) in EBSS. This amount could cover the entire surface of the culture without overflow into the culture medium underneath. Control cultures were treated with an equivalent volume of the EBSS.

For exposure of submerged cultures, the arsenic stock solution was added directly into MEM to obtain desired concentrations. Equivalent volume of the MEM was added into control cultures without arsenic amendment. After exposure, the cultures were washed three times with balanced salt solution to remove any residual arsenic before further "work-up."

MORPHOLOGICAL EVALUATION

Morphology of the fibroblasts and submerged keratinocyte cultures was evaluated by Leitz Diavert phase contrast microscope.

Cultures, grown submerged on the membranes were fixed in neutral buffered formalin (NBF), stained with hematoxylin/eosin and mounted with Permount® on the glass-slide for observation by light microscope (Zeiss). This is usually performed as a quality control procedure before cultures are "lifted" to the air-liquid interface. Artificial skin and submerged cultures, grown on plastic and detached by Dispase® treatment, were fixed in NBF, processed for paraffin embedding and cross-sectioned for subsequent staining.

After 6 and 24 hrs of exposure the cultures were rinsed in three changes of PBS, fixed overnight in Karnovsky solution (Karnovsky, 1965), rinsed with cacodilate buffer (0.1 M, pH 7.4 containing 0.3 M sucrose), postfixed with 2% osmium tetroxide, rinsed in cacodylate buffer, stained with 2% uranyl acetate, rinsed in distilled water, dehydrated in a series of graded ethanols and embedded in Epon. Thin sections were post stained with uranyl acetate and lead citrate.

CYTOTOXICITY EVALUATION

Permeability Assay: Water (Solvent)

Alterations in solvent (water) permeability as a result of exposure to different arsenic concentrations and associated effects were evaluated by the technology previously developed in this laboratory (Pu et al., 1995). The artificial skin was sandwiched between two pieces of Puropore® membranes and inserted into the Teflon® permeation cell, which has cross-sectional area of 0.32 cm^2, and then connected to a Manostat Cassette Pump® (Figure 2.2). The solution (EBSS or NaCL at 30°C) was pumped through the bottom compartment of the permeation cell at a flow rate of 2 ml/h and then collected fractionally at 1-h intervals.

Permeation of tritiated water ^3H_2O, 165500 cpm/ml (New England Nuclear, Boston, MA) was used to evaluate the degree of barrier function for every batch of produced cultures. Radioactivity in 0.2 ml aliquots from the collected samples was counted in a Packard 1500 Tricarb liquid scintillation spectrometer. The permeation constant ($K_{p,w}$) was calculated from the steady-state flux rate as previously described (Pu et al., 1995). Cultures with permeability coefficient of $22\text{-}30\times10^{-3}$ cm/h were used for experiments.

Permeability Assay: Arsenic (Solute)

Uptake of arsenic by artificial skin was determined by applying 400 μL of solution with a given arsenic concentration on top of the cultures in the permeation cell followed by measuring the volume remaining at the end of the experimental run (typically 6 hours). Control cultures were treated with an equivalent volume of the NaCl and no arsenic. The cell then was covered with parafilm to prevent evaporation of the "donor" solution.

Arsenic, penetrating through the artificial skin was collected in a stream of 0.85% NaCL flowing at a rate of 2ml per hour. Samples were collected at 1 hr intervals for up to 6 hours. Arsenic

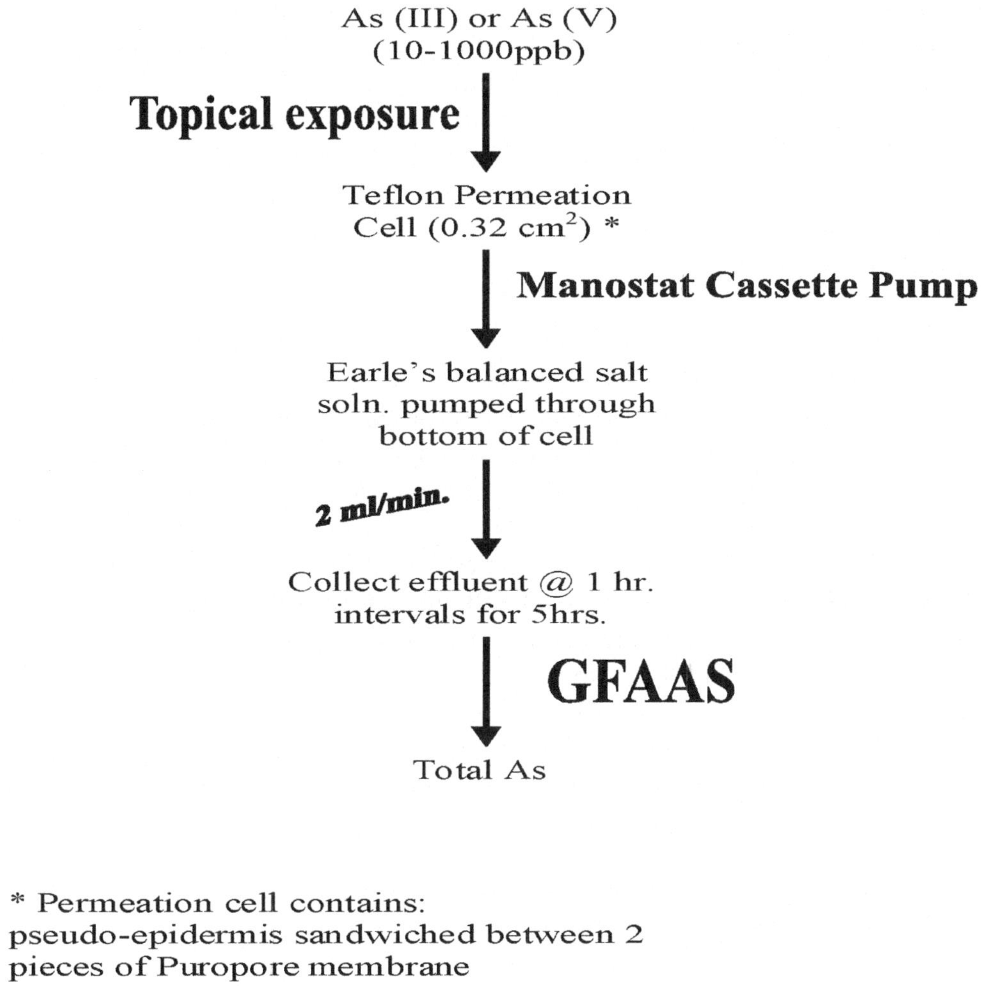

Figure 2.2 Schematic diagram illustrating method for permeability assay technique

concentrations in aliquots of the samples were analyzed by graphite furnace atomic absorption spectrometry (GFAAS; Perkin-Elmer model 4100ZL) equipped with L'vov platform for background correction and electrodeless discharge lamp to increase the sensitivity. All samples were analyzed twice and 0.005 mg of $Pb(NO_3)_2$ and 0.003 mg of $Mg(NO_3)_2$ were added into the blank, standards and samples as a matrix modifier in the GFAAS method used in this study. The introduction of all the samples into the furnace was done using an autosampler system. The sensitivity of the GFAAS was $<10^{-10}$ g arsenic (about 0.12 absorption units for 50 μg/L when 20 μL sample aliquot was used).

Lactate Dehydrogenase (LDH) Assay

This test measures the release of the LDH enzyme into the medium which reflects the cellular membrane integrity. The assay is based on the reduction of nicotin amide adenine dinucleotide (NAD) by the action of LDH and the associated color changes were monitored during the conversion of tetrazolium salt into the formazan product. The optical measurement of the color produced is directly proportional to the number of lysed cells. This method has been routinely used to measure viability of cells exposed to injuring agent (Decker and Lohmann-Matthes, 1988). The LDH released from artificial skin was measured in aliquots of the effluent obtained in permeability studies as well as in submerged culture media after exposure to arsenic by the In Vitro Toxicology Assay Kit for Lactate Dehydrogenase™ (Sigma).

MTS Cell Proliferation Assay

This method is used for determining the amount of viable cells in proliferative or cytotoxicity assays. It is based on color development of the MTS tetrazolium compound [3-(4,5-dimethylthiozol-2-yl)-5-3-carboxymethoxyphenyl)-2-(4-sulfophenyl)-2h-tetrazolium] (Owen's reagent) resulting from the formation of purple-colored formazan product soluble in tissue culture medium. The color formation is due to dehydrogenase enzyme present in metabolically active viable cells and provides an index of cell viability/cytotoxicity (Abe and Matsuki, 2000). This assay has been accepted as a standard and accurate procedure for the measurement of quantitative effects of damaging agents on various types of cells in culture. We evaluated submerged cultures of HK because artificial skin is not permeable to the Owen's reagent. After exposure to arsenic, 200 μL of the assay mixture (MTS/PMS solution, Celltiter 96 Aqueous Non-Radioactive Cell Proliferation Assay Kit™ (Promega) was added to each well. Plates were placed back into the incubator for 4 hours for maximum color development. The reduction of MTS in the presence of cellular dehydrogenases formed formazan which was soluble in tissue culture medium. The absorbance of the complex was measured in a microtiter plate reader at 490nm (Cory et al., 1991).

EVALUATION OF TOXIC EFFECTS OF ARSENIC ON MACROMOLECULAR BIOSYNTHESIS

Artificial skin was topically exposed to 10-1000 μg/mL of arsenic for 0.5-24 hr followed by a gentle rinse of cultures with EBSS to remove residual arsenic. Then the cultures were transferred onto glass fiber filters saturated with MEM containing 2 μCi/mL of [^{3}H]-thymidine (specific activity 80 μCi/mmole), 1 μCi [^{3}H]-uridine (specific activity 26 μCi/mmole) or 1.5 μCi of [^{14}C]-leucine (specific activity 348 μCi/mmole) (all from Amersham-Pharmacia) for 2 hr. The cultures were then rinsed with cold EBSS, and the rates of DNA, RNA, and protein synthesis were estimated from the incorporation of the radiolabeled precursors into cold 5%TCA insoluble material. Precipitate of protein and DNA/RNA was washed two times with cold TCA and lipids were extracted with potassium acetate/absolute ethanol, followed by incubation in absolute ethanol (15 min at 60°C). The precipitate containing DNA, RNA and proteins was dissolved in 0.25N NaOH at 60°C for 30 min. DNA content was measured by the modification of the method of Setaro and Morley (1976) using DABA (diaminobenzoic acid dihydrochloride, Aldrich), which is more selective for DNA since it reacts only with deoxyribose liberated from DNA by acid hydrolysis. This procedure essentially completely degrades RNA.

Aliquots of each sample and standard samples of calf thymus DNA were dried overnight at 60°C. The next day, 100 μL of a 400 mg/ml solution of freshly prepared DABA was added to each tube and incubated for 30 min at 60°C water bath. Two ml of 1N HCL were added to each tube to stop the reaction. The reaction yields a product with a fluorescent wavelength maximum at 520 nm, which is specific for deoxyribose. Each tube was read at an excitation wavelength of 410 nm and an emission wavelength of 505 nm using a Perkin-Elmer fluorescence spectrophotometer.

Incorporation of [^{3}H]-uridine into TCA-insoluble RNA was calculated per μg of total TCA-precipitable protein to minimize the error resulting from the procedural step involving dissolving of cold TCA insoluble precipitate in hot NaOH that is known to degrade almost completely RNA sparing TCA-insoluble DNA. Since the radioactivity counting was performed in the dissolved precipitate without re-precipitation, the entire [^{3}H]-uridine incorporation was thereby accounted for and represented incorporation into RNA on the basis of the precursor specificity.

Protein content was assayed by the Lowery method (Lowry et al., 1951). Incorporation of radiolabelled precursors was expressed as cpm per μg of DNA or protein.

EVALUATION OF EFFECTS OF ARSENIC ON PROGRAMMED CELL DEATH

DNA Fragmentation Assay (TUNEL Assay)

Apoptotic changes were detected using the TUNEL assay (Apop-Tek, Intragen) in paraffin sections of submerged and lifted cultures exposed to As(V) and As(III) to document DNA fragmentation in exposed cells *in situ*. The assay relies on the labeling of free 3'-OH DNA termini *in situ* with digoxigenin-labeled and unlabeled nucleotides. The nucleotides are enzymatically added to the DNA by TdT enzyme, which catalyzes template independent addition of nucleotide triphosphates to the 3'-OH ends of double-stranded or single-stranded DNA. Labeled DNA is then bound by an anti-digoxigenin antibody that is conjugated to a peroxidase. The bound peroxidase antibody conjugate enzymatically generates permanent stain from chromogenic substances. Localized staining of DNA strand brakes in apoptotic cells was visualized by light microscopy.

Apoptosis Evaluation at the Single Cell Level (COMET Assay)

The assay relies on the formation of comet-like tails in the gel electrophoresis that are produced by DNA denatured as a result of exposure to toxic agents. Undamaged DNA migrates in the gel at a slower rate and remains within the confines of the nucleoid and thus appears as an intact comet head. DNA fragments migrate relatively faster and form a comet-like tail whose fluorescent intensity and shape can be related to the level of damage (Ostling and Johanson, 1984). The extent of the damage can be quantified by measuring the length, shape, intensity of staining, etc of the tail in comparison to the "head" of the comet. Briefly, the Comet assay involves the measurement of resistance of alkyl-labile sites and DNA single strand breaks by the method of Singh et al. (1988). Fully frosted slides (Fisher) were precoated with 100 μL of 0.5% normal melted agarose (Sigma) in phosphate buffered saline (PBS) and allowed to solidify at 4°C for 5-10 min as shown in Figure 2.3. Following exposure of cells to the selected doses of arsenic the cells were trypsinized, washed in ice-cold PBS and made into single cell suspension which is further re-suspended in 0.5% low-melting agarose at 37°C at a concentration of 2-4 $\times 10^4$ cells/ml. One hundred microliters of this suspension were pipetted onto the pre-coated glass slides and allowed to set at 4°C. After that, another 100 μL of the low melting agarose (0.5%) was layered on to the previous cell-containing layer and allowed to solidify. The samples were then placed in freshly prepared lysis solution (100 mM Na_2EDTA, 2.5M NaCl, 10mM Tris, 1% Triton X-100, pH 10, 4°C) for at least 1 h. Following lysis, the slides were drained and incubated in electrophoresis buffer (300 mM NaOH, 1mM Na_2EDTA, pH >13) for 20 min for DNA unwinding, followed by electrophoresis in the same

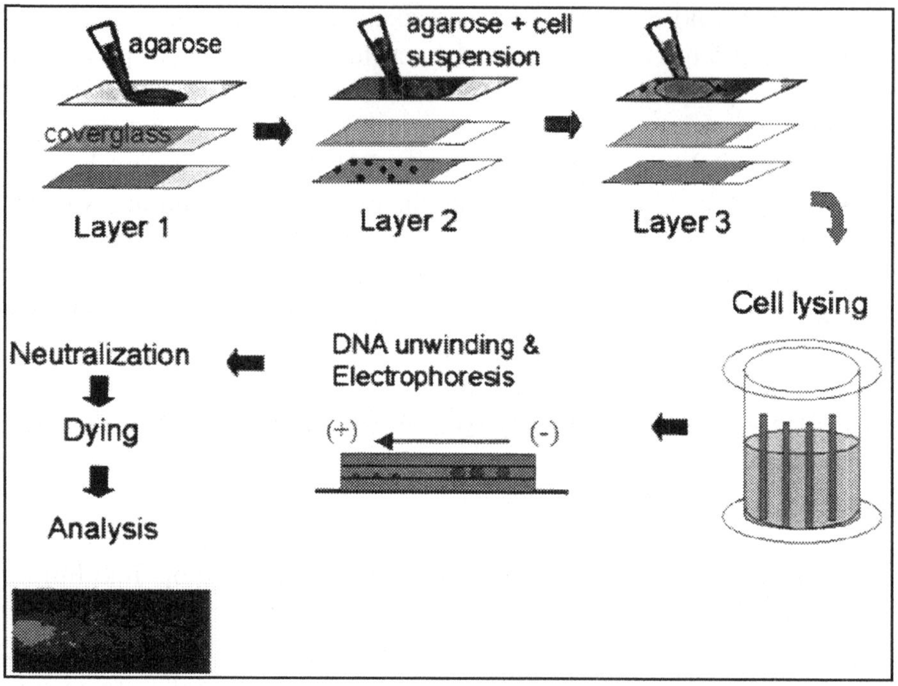

Figure 2.3 Schematic diagram illustrating Comet Assay: A microgel electrophoresis technique

buffer for 20 min (25 V, 300 mA). After electrophoresis the slides were washed three times with neutralization buffer (0.4M Tris, pH 7.4) for 5 min each to remove alkaline buffer. Then slides were briefly immersed into 70% ethanol, dried in room temperature, and stored in PBS saturated box at 4°C until analysis. All procedures were performed either in the dark or at dim yellow light to prevent possible UV damage of DNA.

Subsequently, the cells were stained with 100 μl of either ethidium bromide (20 μl/ml;) or VISTRA green (1:10,000) fluorescent dye prior to image analysis of the comets. DNA tails were visualized using Olympus epifluorescence microscope with 516-560 nm excitation from a 50 W mercury light. When excited, the DNA-bound VISTRA green emits green light whereas ethidium bromide emits red light. In control cells, fluorescence is confined to the nucleus: undamaged DNA is supercoiled and does not migrate out of the nucleus under the influence of an electric current. In damaged cells, the alkaline treatment unwinds the DNA to release fragments that migrate through the nuclear membrane under the influence of the electric field. The negatively charged DNA migrates toward the anode and the length of travel of DNA fragments in the electrophoretic gel reflects the increasing relaxation of the supercoiled DNA molecules correlated with the extent of DNA damage by the toxic agent. Images of comets are captured on film, and quantitative analysis is performed using proprietary software package. Up to fifty comets were randomly selected per a slide, and three slides were used per one experimental dose.

Apoptosis Evaluation on the Chromosome Level (Telomerase Assay)

Telomerase PCR ELISA method have been used to quantify the telomerase activity in cell extracts from submerged cultures of human skin cells. Cell extracts were prepared from HK and

HF and exposed to As(III) (100 and 1000 μg/L for 15 and 30 min). Telomerase activity was determined by the Telomeric Repeat Amplification Protocol (TRAP) assay used for amplification of the telomerase-reaction product. This is a sensitive photometric enzyme immunoassay for the detection of telomerase activity using no radioactive techniques. DNA free RNase was preincubated within cell extracts from the negative control group to completely destroy any expression of telomerase, and, therefore, to insure the specificity of this experiment. For TRAP reaction we used GeneAmp PCR System 9600 Thermal Cycler and DNA Thermal Cycler 480 system. Primers were elongated for 20 min at 25°C, telomerase was inactivated for 5 min at 94°C for 2 cycles and the products were amplified through 94°C, 50°C and 72°C for one 30 sec cycle, then 10 min at 72°C for 33 cycles. Elongation/amplification products from the first step were denatured and hybridized to a digoxigenin-(DIG)-labeled telomeric repeat- specific detection probe. The resulting PCR products are immobilized via biotinylated primer to microtiter plate, followed by ELISA.

Chromosome Analysis

Chromosome preparations were obtained from fibroblast cultures exposed to different concentrations of As(V) and processed for cytogenetic evaluation according to the conventional protocols. Fibroblasts were plated on cover slips at low density (200×10^3). At 24-hr, cells were exposed to different doses of As(V) for 3 hr. Colcemid (0.05 μg/ml) was added for the last hour of incubation. The cells were treated with hypotonic solution, fixed with methanol:acetic acid (3:1, v/v), dried and stained with Giemsa stain. One thousand cells were scored for mitotic index.

Analysis of Toxic Effects of Arsenic on Signal Transduction (Western Blot)

For Western blots, human keratinocytes grown submerged and/or HF were detached from the dish by scraping the cells into lysis buffer (1% Nonidet P-40, 20mM Tris (pH 8.0), 137mM NaCl, 10% glycerol, 1mM Na_3VO_4, 2mM EDTA, 1mM phenylmethylsulfonyl fluoride, 20μM leupeptin, and 0.15 unit/ml aprotinin). Thirty micrograms of the isolated proteins were fractionated using PAGE, the bands were transferred to the PVDF membrane and the western blot performed. The primary HSP70 and p38 antibodies were from Santa Cruz Biotechnology. The secondary antibodies were supplied by Amersham Pharmacia Biotech. The detection of membrane-bound primary antibodies was done by the ECL Plus technique (Amersham Pharmacia Biotech) using Storm 860 Molecular Devices system.

Genotoxicity Experiments

The first factor determined was p53. Expression of p53 protein was visualized using a monoclonal antibody. Artificial human skin was exposed for 1, 6, 24, 48 and 72 hr by topically applying 10, 100 and 500 μg/L of As(III) in EBSS. After exposure to arsenic compounds the cultures were washed three times with PBS to remove any residual arsenic, fixed in NBF and embedded in paraffin. Cross-sections of cultures were used for staining with Monoclonal Mouse Anti-Human p53 Protein (clone DO-7, Dako, CA). Expression of the p53 protein was determined using a monoclonal antibody that recognizes nuclear protein p53. De-paraffinized sections were rehydrated, endogenous peroxidase activity was quenched (5 min) by 3% hydrogen peroxide and rinsed with Tris buffered saline with Tween 20 (TBST). Samples were incubated with monoclonal primary antibody DO-7 (1:200 dilution, 30 min., room temperature), washed with TBST.

Secondary antibodies (biotinylated universal antibodies, 1:200 dilution) were added for 30 min at room temperature. Immunoreaction was visualized using avidin/biotin system and peroxidase-labeled streptavidin (DAKO) incubated for 30 min at room temperature. Staining was completed after incubation with 3-3' diaminobenzidine (DAB) substrate Chromagen (5 min), followed by rinsing with water and counterstaining with hematoxylin (15 sec). Samples were dehydrated and mounted with Permount. Stained slides revealed a brown-colored precipitate at the antigen sites.

INHIBITION OF ARSENIC EFFECT

The MAP kinase inhibitors SB 203580, PD 98059, SB 202190 and SB 202474, the negative control for MAP kinase inhibition studies were purchased from CalBiochem, CA. COX-2 inhibitors NS-398, Niflumic acid and nimesulide were from Cayman Chemical Co, Ann Arbor, MI. The cells were exposed to inhibitors for 1 hr prior to their transfer to the medium containing As(III) in the final concentration for 2, 3 and 24 hr as required by the experiment. The concentration of MAPK inhibitors was 10 μM and that of COX-2 inhibitors 50 μM. Following exposure to inhibitors and/or As(III) the cells were rinsed thoroughly and processed for immunoperoxidase staining or phase contrast microscopy.

The primary antibodies against MAP kinases, namely, extracellular regulated kinase (ERK), c-jun terminal kinase (JNK), p38 and the biotinylated secondary antibodies were from Santa Cruz Biotechnology. The detection of membrane-bound primary antibodies was done by immunoperoxidase staining.

Diaminobenzidine tetrahydrochloride (DAB) was used to localize peroxidase in tissue sections. DAB produces a gray/black precipitate in the presence of some divalent cations. Sections or monolayer cultures were first blocked, incubated with the primary antibody, then biotinylated secondary antibody were added, followed by the addition of the HRP-Streptavidin complex. The target protein was visualized by incubation with the peroxidase substrate, using DAB as the chromagen.

Artificial skin was topically exposed to 5, 10 or 100 μg/L of As(III), rinsed in PBS and embedded in paraffin blocks according to standard procedures. Tissue sections (3-4 microns thick) were deparaffinized in xylene using two changes of 5 minutes each, hydrated through graded alcohols and washed in deionized H_2O for 1 minute on stir plate. Since certain antigenic determinants are masked by formalin fixation and paraffin embedding, antigen unmasking was performed by heat treatment. Slides were placed in a container, covered with 10 mM sodium citrate buffer, pH 6.0 and heated at 95°C for 5 minutes. Fresh buffer was added and slides were heated at 95°C for additional 5 minutes, placed in the buffer to cool down for 15 minutes and washed in deionized H_2O. Excess liquid from slides was aspirated and a blocking reagent (1.5% goat serum) was added to each slide for 30 min followed by addition of 100 μl of an appropriate primary antibody (1:50 dilution). Slides were incubated for 2 hours at room temperature, washed in PBS twice for 3 minutes each on a stir plate. Excess liquid from slides was aspirated, slides were incubated for 30 minutes in 100 μl of biotinylated secondary antibody and washed as above.

Finally the slides were incubated for 30 min with horseradish peroxidase (HRP)-streptavidin complex, washed as above and HRP substrate was added for 4-8 minutes (until desired stain intensity developed). Slides were then washed with deionized H_2O, counterstained with Gill hematoxylin, washed with H_2O, dehydrated, mounted in Permount and observed by light microscopy.

Human keratinocytes and fibroblasts were plated on sterile coverslips ($100 \times 10^3/cm^2$) and grown for 24 and 48 hr. Cells were washed with PBS and exposed to 5–1000 µg/L of As(III) for 2, 3 and 24 hrs. Exposure for short time (up to 3 hr) was done in the serum-free medium in order to minimize the neutralizing effects of serum on the inhibitors. For longer exposures a complete growth medium was used. After exposure the cells were rinsed 3 times in PBS, pH 7.2, fixed for 2 minutes in cold acetone, air dried and washed in three changes of PBS. Background staining was eliminated or significantly reduced by addition of the blocking reagent. All subsequent staining was performed as written above for artificial skin.

ARSENATE REDUCTION BY AMINO ACIDS

The amino acids used in the study included L-glutamine ($C_5H_{10}N_2O_3$), L-methionine ($C_5H_{11}NO_2S$), L-cysteine ($C_3H_7NO_2S$), and homocysteine ($C_4H_9NO_2S$). Glutamine, methionine, and cysteine were obtained from Fisher; and homocysteine was obtained from Sigma. The preparation of solutions was adapted and modified from the method described by Zhitkovich, Shrager, and Messer (2000). Starting solutions consisted of 4mg/L of amino acid and 2mg/L As(V), leading to a molar ratio of amino acid:As being 2:1. As(V) solution was obtained from Fisher. Controls were prepared with the same As(V) concentration, and with no amino acid present. All starting solutions were prepared in de-ionized water. The reactions were performed at body temperature (37°C) and samples were collected every 24 hours up to 7 days.

At the end of exposure, solutions were passed through a column packed with anion exchange resin (AG1-X8) to separate As(V), retained in the column, from As(III) which remained in the effluent. The As(V) retained by the anion exchange column was eluted using 0.1M HCl. Collected samples were diluted 21 times, and the As(III) and As(V) fractions collected were determined using inductively coupled plasma mass spectroscopy (ICPMS). Percent reduction was calculated using the following equation:

$$\%\ \text{reduction} = \frac{[\text{As(III) (fractions collected from speciation)}] \times 100}{[\text{starting As(V) concentration}]}$$

As(III) fractions collected from speciation are analogous to As(V) fractions collected minus starting As(V) concentrations.

The effect of amino acid concentration on the non-enzymatic reduction of As(V) was studied by increasing and decreasing the amount of amino acid by a factor of 10. To study the effect of increased amino acid concentration on As(V) reduction, starting solutions consisted of 40 mg/L of amino acid and 2 mg/L As(V), leading to a molar ratio 20:1. Solutions were also prepared to examine the effect of decreased amino acid concentrations on As(V) reduction. In this case, solutions consisted of 0.4 mg/L of amino acid and 2 mg/L As(V), leading to a molar ratio of 0.2:1. In this part of the experiment, only the starting solutions changed. Again, controls were prepared with the same As(V) concentration, and with no amino acid present. All starting solutions were prepared in de-ionized water. Again, the reactions were performed at 37°C and samples were taken every 24 hours up to 7 days. The separation and analysis of As(III) and As(V) followed the method as described above.

QUALITY ASSURANCE PROGRAM

The Quality Assurance Program (QAP) for the experiments and analysis of arsenic in the samples followed closely those already developed for the study of metals at ultra-trace levels. Because of improvements in sample collection and handling and in the instrumental detection limits, there has been a leap in our understanding of the environmental chemistry of trace metals in many ecosystems. This laboratory has been at the forefront of these analytical developments.

A prerequisite for speciation studies is the collection and processing of samples uncontaminated with the element of interest or other potentially competing metals/metalloids. Most of the previous speciation experiments in natural waters were jeopardized or rendered meaningless by inadvertent contamination and inadequate processing techniques. Great care was used in these studies to ensure that a high degree of sample integrity is maintained during the experiments and subsequent collection and analysis of samples. Scrupulous cleanliness and highly sensitive analytical techniques were employed in speciation studies given that only a fraction of the total concentrations (of a few ng L^{-1}) were analyzed (Beuabien et al., 1993). Most of the experiments as well as sample collection, handling and analysis methods were performed using ultra-clean laboratory techniques in a Class 100 laboratory. The methods for decontaminating the bottles and labware followed the nine step procedure described by Nriagu et al. (1993).

Highest quality reagents and chemicals were used throughout the study. Extreme care was taken to minimize the risk of contamination during all experiments. Although experiments with As(III) were done in a nitrogen atmosphere, the final solutions were analyzed for both As(III) and As(V) to ascertain the degree of oxidation of arsenite during each experiment. A careful blanking scheme was performed to estimate the level of arsenic contamination from every facet of the experiment and during sample analysis. These blanks were also used as diagnostic tools and to generate method detection limits. The blanking scheme used in the study included the following:

Sample Type	Experimental replicates	Analyte spike (sample matrix)	Analyte spike (blank matrix)	Bottle and labware blank	Reagent blank
QAP Frequency	15-20%	10-15%	5-10%	5-10%	2-5%

Arsenic concentrations were determined by GFAAS or ion chromatography, and the speciation of metals was done using a column separation technique, as described above. The following batch analysis QAP was implemented to assess experimental and/or instrumental performance, recovery, precision, and accuracy.

QAP Step	Frequency
Instrumental analysis	
Blank levels	Before each experimental batch
Stability	Before each experimental batch
Sensitivity	Before each experimental batch
Resolution	Before each experimental batch
Interference check	Once per week

Blank levels during run	
Calibration blank	Once per experimental batch
Check blanks	Four per experimental batch
Memory check	One per experimental batch
Recovery	
Lab analyte spike, blank matrix	One per experimental batch
Lab analyte spike, sample matrix	Two per experimental batch
Internal standard, 2-metals	Two per experimental batch
Precision	
Replicate sample acquisitions	Four per experimental batch
Lab sample replicates (within batch)	Two per experimental batch
Lab sample replicates (diff. batches)	20%
Accuracy	
Standard Ref. Material	One per experimental batch
Lab control sample	One per experimental batch

Data from the various experiments was subjected to multidimensional scaling, regression, cluster, factor analyses, and ANOVA and were used to test models of multivariate and nonlinear relationships. Most of the data analysis was done using off-the-shelf software.

DATA ANALYSIS

The incorporation of radioactive precursors was analyzed by determination of the mean and standard deviation of the data. For the experiments, each set of control and exposed cultures was carried out in triplicate wells. Each well was assayed in triplicate for cytotoxicity and viability (LDH and MTS assays) when submerged cultures were evaluated. Statistical significance was determined using the student T-test. Permeation experiments were repeated in triplicate to derive the mean and error ranges for the serial samples.

Images of comets were evaluated by a computer-assisted image analysis software package. Statistical analysis of the difference between means was carried out using a one-way ANOVA, and results were considered statistically significant at the 5% level.

The cutoff point for positive telomerase activity was set at a difference greater than 0.2 OD units in the optical densities, A_{450} nm - A_{690} nm, and any results with lower than 0.2 units were considered to be negative. Each treatment was performed in duplicate and the average value was used.

CHAPTER 3
RESULTS

MORPHOLOGICAL STUDIES

Submerged Cultures

Human epidermal keratinocytes were exposed to 10, 100, 250, 500 and 1000 μg/L of As(V) or As(III) beginning on the second day of growth (24 hr old cultures) and examined at the times indicated.

Exposure to Arsenate

Cell samples from control cultures formed characteristic monolayers of epithelial growth featuring round shaped cells with well-defined intercellular spaces (Figure 3.1A). On the other hand, gross morphological changes, detectable by phase contrast microscopy during the following four days of growth in the presence of As(V), were observed in submerged cultures of HK. The most pronounced changes at high As(V) doses (500 and 1000 μg/L), detectable by phase contrast microscopy by the third day of exposure, involved the stratification and differentiation of keratinocytes that contrasted sharply with the morphology of unexposed cells (Figure 3.1A) or those exposed to 10 μg/L (Figure 3.1B) and 100 μg/L (Figure 3.1C) of As(V). In many areas, flattened cells with thick cell membranes were seen above small cells without clear margins (Figures 3.1 E-F). Morphology of cultures exposed to 250 μg/L showed similar changes although to a much lesser extent (Figure 3.1D). The effects of arsenic on primary submerged cultures of human keratinocytes exposed to 500 and 1000 μg/L of As(V) were strikingly similar to the gross morphological appearance of clinical skin lesions reported from *in vivo* exposure to high arsenic levels in endemic areas (Figure 3.2).

The morphology of the cultures exposed to different levels of As(V) for the 7 days did not change significantly from those described above. However, at the lower doses (100 and 250 μg/L) the exposed cultures developed increased stratification.

Analysis of paraffin sections of these differentiating and stratifying cultures grown on plastic dishes submerged and exposed to As(V) for 3 hr revealed a generalized disorganization of the basal layers accompanied by a significant increase in mitotic activity and pronounced pyknosis (Figure 3.3). Control cultures showed cuboidal basal cells arranged in a regular columnar manner. At lower doses of As(V), the submerged cultures either did not show any morphological abnormalities (10 μg/L; Figure 3.3B) or only focal disruption of the basal layer with occasional karyopyknotic cells at 100 and 250 μg/L (Figure 3.3 C, D). Changes in morphology progressively increased with higher doses of As(V) eventually leading to disintegration of the basal layer with either markedly pyknotic or enlarged nuclei and abundant mitotic figures (Figure 3.3 E-F).

Exposure to Arsenite

Cultures of HK grown submerged and exposed to As(III) displayed morphological changes after 36 hr (Figure 3.4). At 10 μg/L, a number of large vacuolated cells with irregular nuclear outlines and dispersed chromatin appeared (Figure 3.4 B, C). In some areas, irregularities

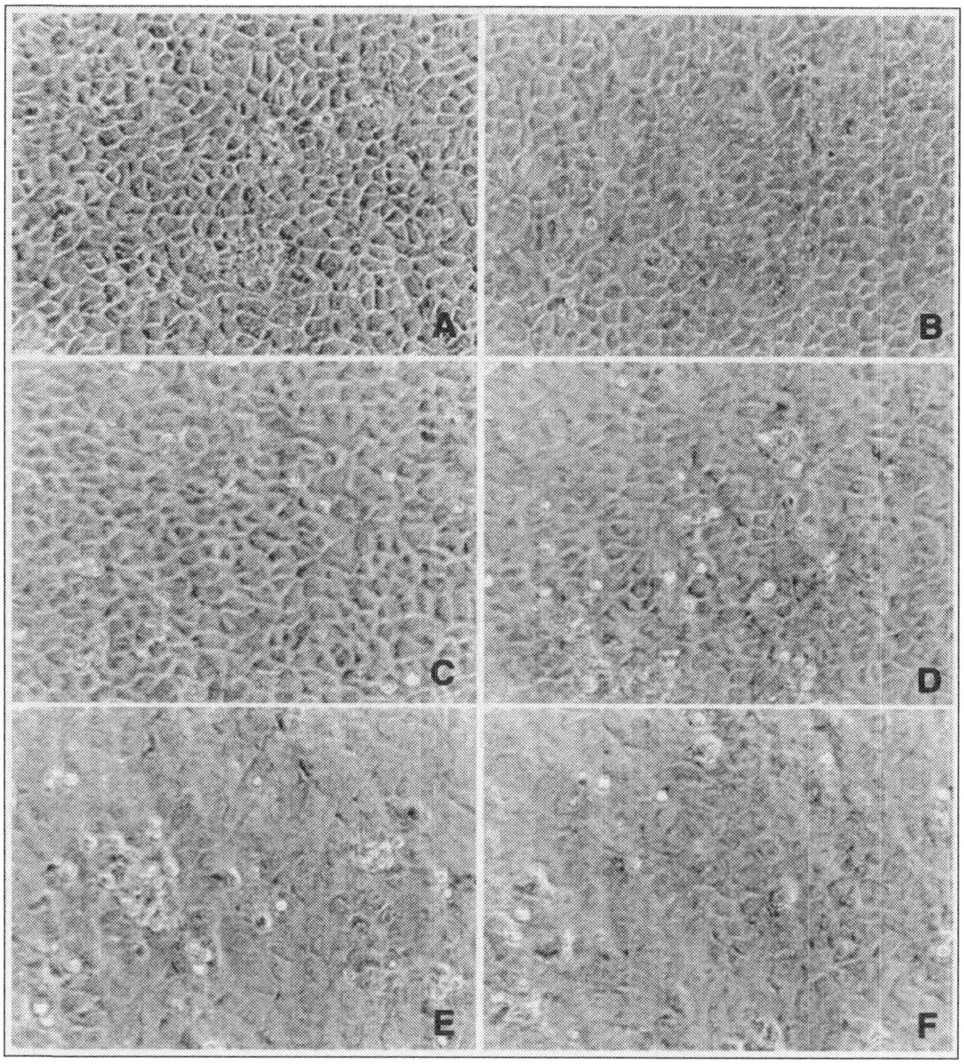

Figure 3.1 Phase contrast micrograph of submerged cultures of human keratinocytes showing morphological changes produced by exposure to As(V). A: control; B–F: three days after the cultures were exposed to 10, 100, 250, 500 and 1000 μ/L As(V), respectively. Note pronounced changes involved striking stratification and differentiation of keratinocytes. Magnification ×230.

were noted in the arrangement of cells and appearance of scattered foci of stratified cells (Figure 3.4C). Similar, but even more pronounced changes, could be seen in cultures exposed to 500 μg/L (Figure 3.4D). At 1000 μg/L, the number of live cell dramatically decreased with the appearance of grossly enlarged, enucleated, epithelial-like cells (Figure 3.4 D, F). These experiments clearly show cellular effects at 10 μg/L, the maximum contaminant level (MCL) for arsenic recently announced by the USEPA (2001).

Phase contrast studies of cells exposed to As(III) for 7 days showed the progressive nature of disintegration of ordered architecture of cellular growth coordination (Figure 3.5C) with deterioration of uniform culture into clusters of cells becoming progressively smaller in size during

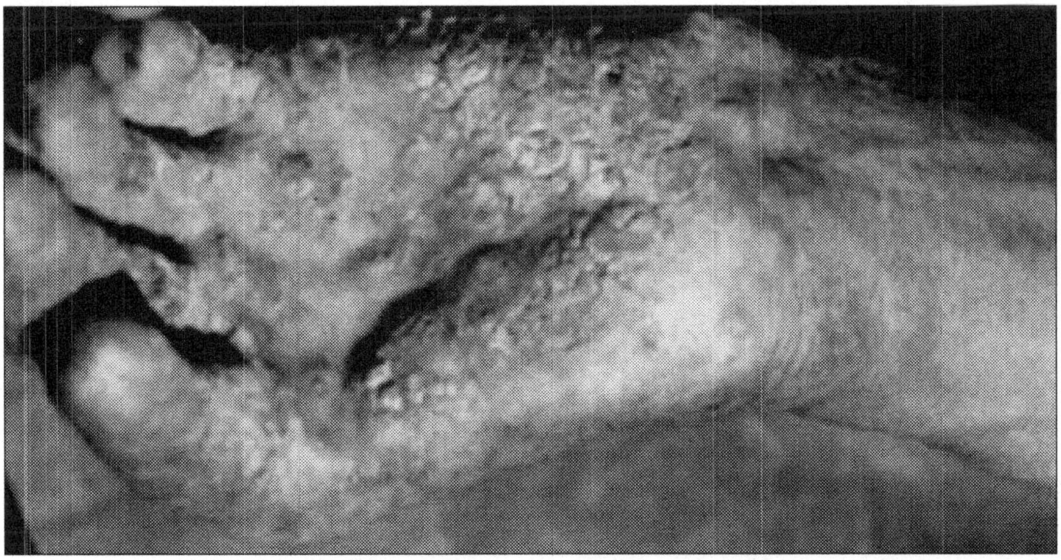

Figure 3.2 Human skin, exposed to high arsenic concentrations in an endemic area (inner Mongolia; left arm skin of male farmer) (Professional courtesy of Dr. Chen).

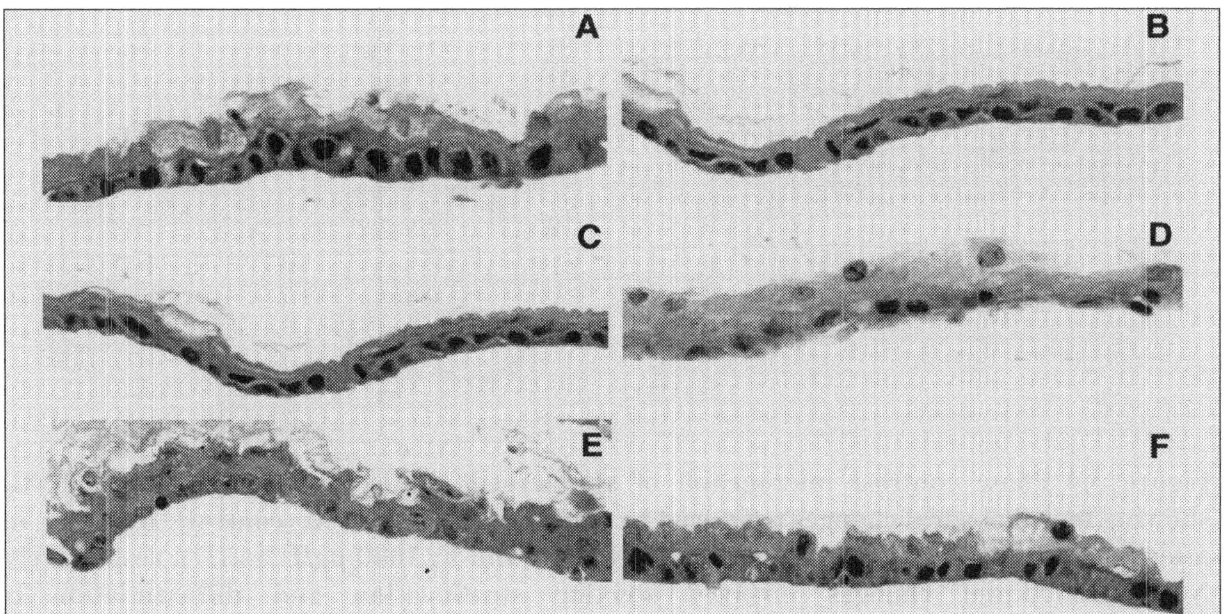

Figure 3.3 Photomicrographs showing dose-dependent changes in submerged cultures exposed to As(V) for 3 hr. A: control; B–F: exposure to 10, 100, 250, 500 and 1000 µg/L As(V), respectively. Note pyknotic nuclei (C, F) and cells with mitotic figures (E). Magnification ×590.

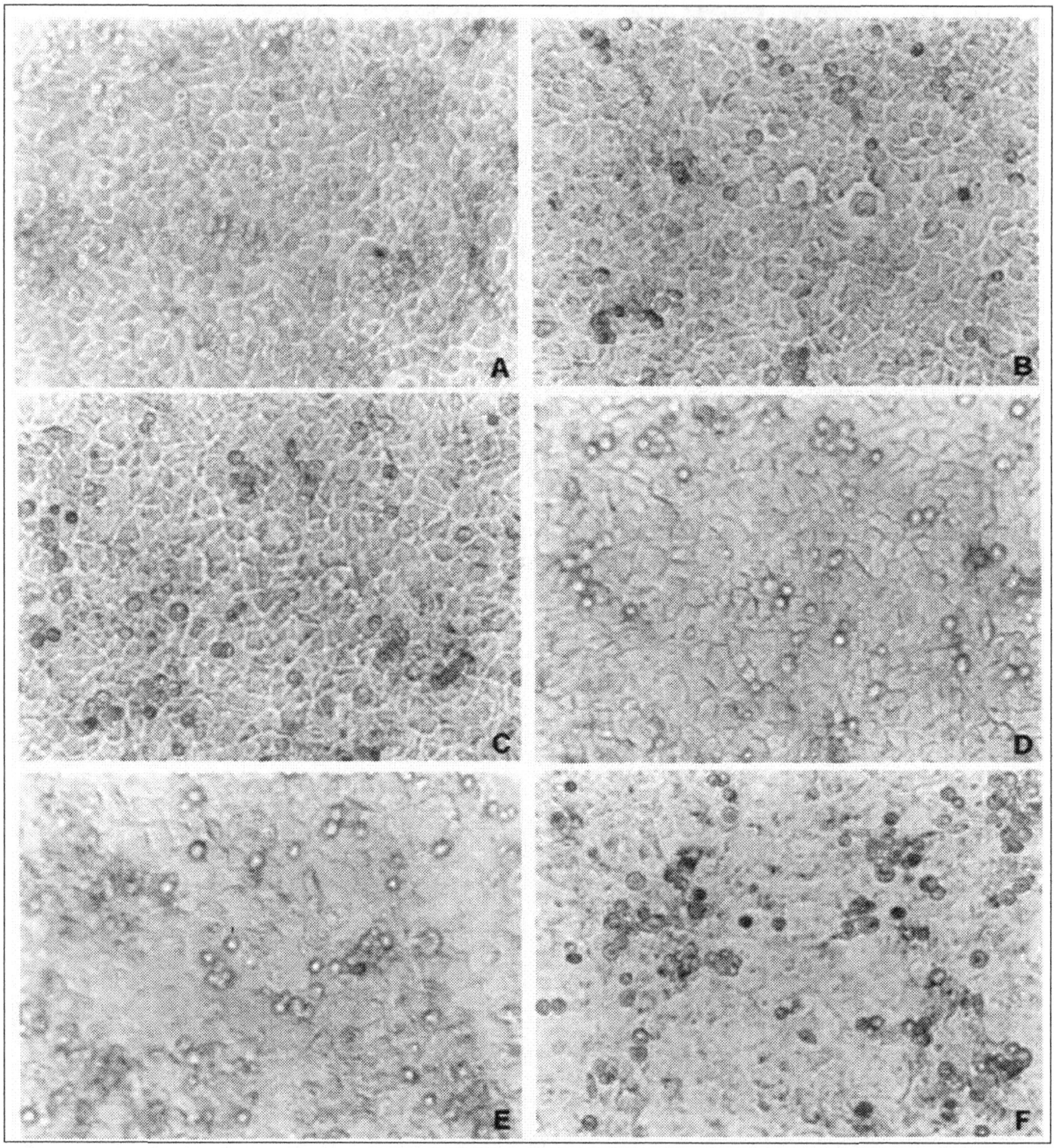

Figure 3.4 Phase contrast micrograph of submerged cultures of human keratinocytes showing morphological changes produced by exposure to As(III). A: control; B–C: 36 hr after the cultures were exposed to 10 μg/L; D: 500 and F: 1000 μg/L As(III), respectively. Note pronounced changes involved striking stratification and differentiation of keratinocytes. Magnification ×230.

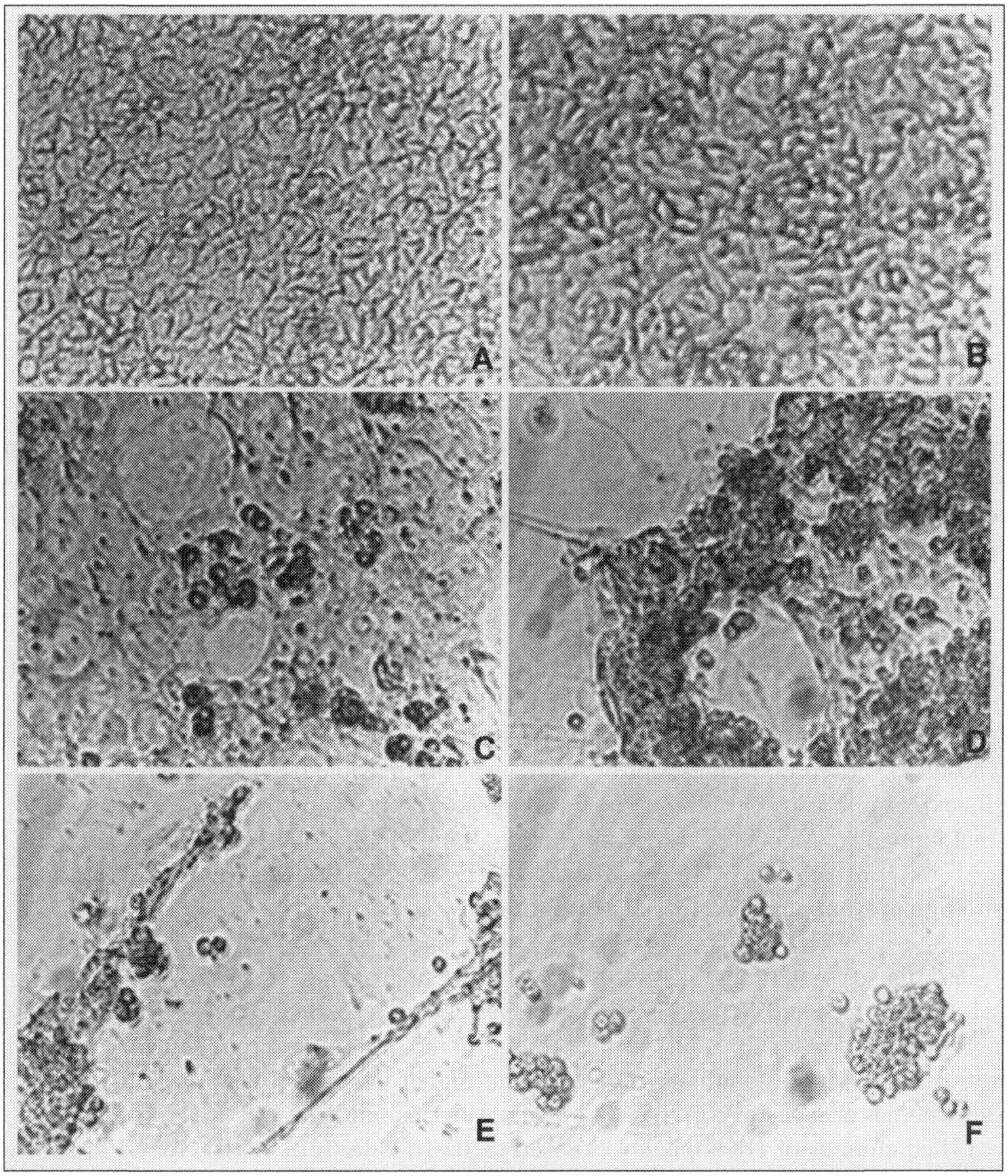

Figure 3.5 Phase contrast micrograph of submerged cultures of human keratinocytes showing morphological changes produced by exposure to As(III). A: control; B–F: seven days after the cultures were exposed to 10, 100, 250, 500 and 1000 µg/L As(III), respectively. Note irreversibly injured cells in D–F. Magnification ×230.

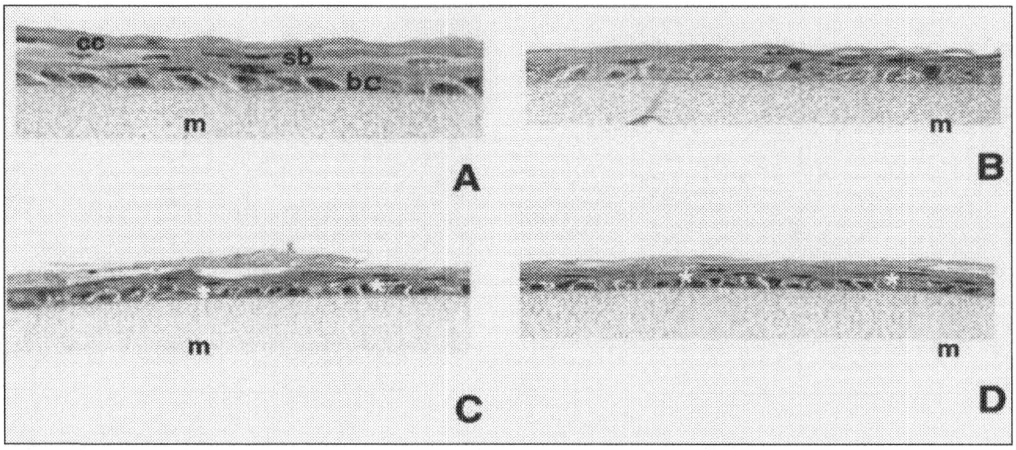

Figure 3.6 Photomicrographs showing dose-dependent changes in artificial skin topically exposed to As(V) for 24 hr. A: control; B–D: exposure to 10, 500 and 1000 μg/L As(V), respectively. cc = cornified cells, sb = suprabasal layers, bc = basal cells, m = membrane. Note pyknotic nuclei (*) in C and D. Magnification ×590.

exposure to 250 μg/L of As(III) (Figure 3.5D). Cells treated with the highest concentration of As(III) showed advanced degenerative changes (Figure 3.5 E-F).

Analysis of paraffin sections of differentiating and stratifying cultures grown submerged and exposed to various concentrations of As(III) was not feasible due to extensive deterioration of the cells exposed to concentrations above 100 μg/L (data not shown). These experiments further show that human keratinocytes are sensitive to As(III) with the effects dose being <10 μg/L.

Morphological Changes in Artificial Human Skin

Paraffin sections of cultures lifted for 14 days and topically exposed to 10 - 1000 μg/L of As(V) for 24 hr revealed noticeable changes even at the lowest dose. These included pyknotic changes, shrinkage of cells and loss of the ordered columnar arrangement of basal cells (Figure 3.6 B-D). The differentiated cell layers at higher doses (500 and 1000 μg/L) were more compact whereas changes in cornified cell layers were insignificant.

Lifted cultures of HK topically exposed to 10-1000 μg/L of As(III) for 24 hr showed that basal cells were the first to become affected. The damage was generally similar to that observed in As(V)-treated cells with marked ballooning of basal cells and an increase in mitotic activity even at the10 μg/L dose. Disruption of intercellular adhesions and disorderly differentiation of early spinous cells were seen. In general, morphological changes in As(III)-treated cells resembled those observed with As(V) exposure except that changes at the doses were accompanied by progressive loss of cells that left only a few cornified layers. These morphological studies with artificial human skin culture thus give results that are similar to those obtained with normal human skins, and show the toxic effect dose for both As(V) and As(III) to be <10 μg/L.

ULTRASTRUCTURAL CHANGES IN ARTIFICIAL SKIN EXPOSED TO AS(III) AND AS(V)

Electron Microscopy Studies

The ultrastructural morphology of cells exposed to 100 µg/L of As(V) for 6 hrs was similar to the morphology of cells in untreated cultures. Typical epidermal cell organelles such as desmosomes, tonofilaments and keratohyalin-like granules could still be observed. Cornified layers (as many as 15 layers could be easily counted) with detectable cornified envelopes could be seen in these cultures as well as in controls (Figure 3.7A). After 24 hr of exposure to 100 µg/L of As(V), the cultures display thick and compact cornified layers, emptied lipid vacuoles, thickening and some disorganization of tonofilament bundles as well as enlarged nuclei with margination of nuclear chromatin along the nuclear membrane. Prominent keratohyalin-like granules containing high molecular weight proteins were seen indicating that an undisturbed process of terminal differentiation of keratinocytes had taken place at this dose of exposure to As(V) (Figure 3.7 B, C). Mitochondria and desmosomes did not appear to be affected in the late spinous cells, but in the basal area there were fewer desmosomes, and occasional degenerated mitochondria could been detected (Figure 3.7D).

On the other hand, cultures exposed to As(III) revealed markedly higher toxicity than As(V) in terms of ultrastructural changes beginning with exposures to10 µg/L for 6 hr. This was seen in a tendency toward reduced adhesion between cells and in the enlargement of the intercellular spaces containing microvilli (Figure 3.8A). Some cells showed an enhanced concentration of mitochondria crowded around the nuclei (Figure 3.8B). At higher doses (exposures to 100 µg/L for 6 hr), a reduction in the amount of cornified cell layers was noticed, as well as disorganization of nuclear chromatin, enlargement of the nuclei, high number of degenerated mitochondria, thickening, and shortening of tonofilaments (Figure 3.8C). At 500 µg/L exposure for 6 hr, nuclear deterioration in basal cells appeared to be similar to that found at 100 µg/L exposures; however, in late spinous cells, the nuclei were completely disorganized suggesting apoptotic changes (Figure 3.8D).

Longer exposures (24 hr) to 10 µg/L of As(III) did not lead to total disorganization of cellular ultrastructure but resulted in a mixture of normal looking and largely destroyed, ballooned cells. However, at higher concentrations of As(III) (100 µg/L) we observed complete loss of cornified cells in some parts of the cultures, further degeneration of nuclear structure, and a reduced number of desmosomes (Figure 3.9A). At 500 µg/L we observed irregular clumping of nuclear chromatin, vacuolization of early spinous cells, and progressive deterioration of nuclear structure in many basal cells (Figure 3.9 B, C). Exposure to 1000 µg/L As(III) for 6 and 24 hr resulted in completely erased cellular structure making ultrastructural studies impossible whereas the exposure to 1000 µg/L As(V) was found to cause consistent and uniform differentiation process including stratification and cornification of the artificial skin.

Electron microscopy observations on skin cultures exposed to As(III) and As(V) revealed concentration and time-related ultrastructural damages that were consistent with the morphological changes observed by phase-contrast microscopy and in cross-sections of artificial skin. The relationships suggest that the no-effect dose is below 10 µg/L, the recently promulgated national MCL for arsenic.

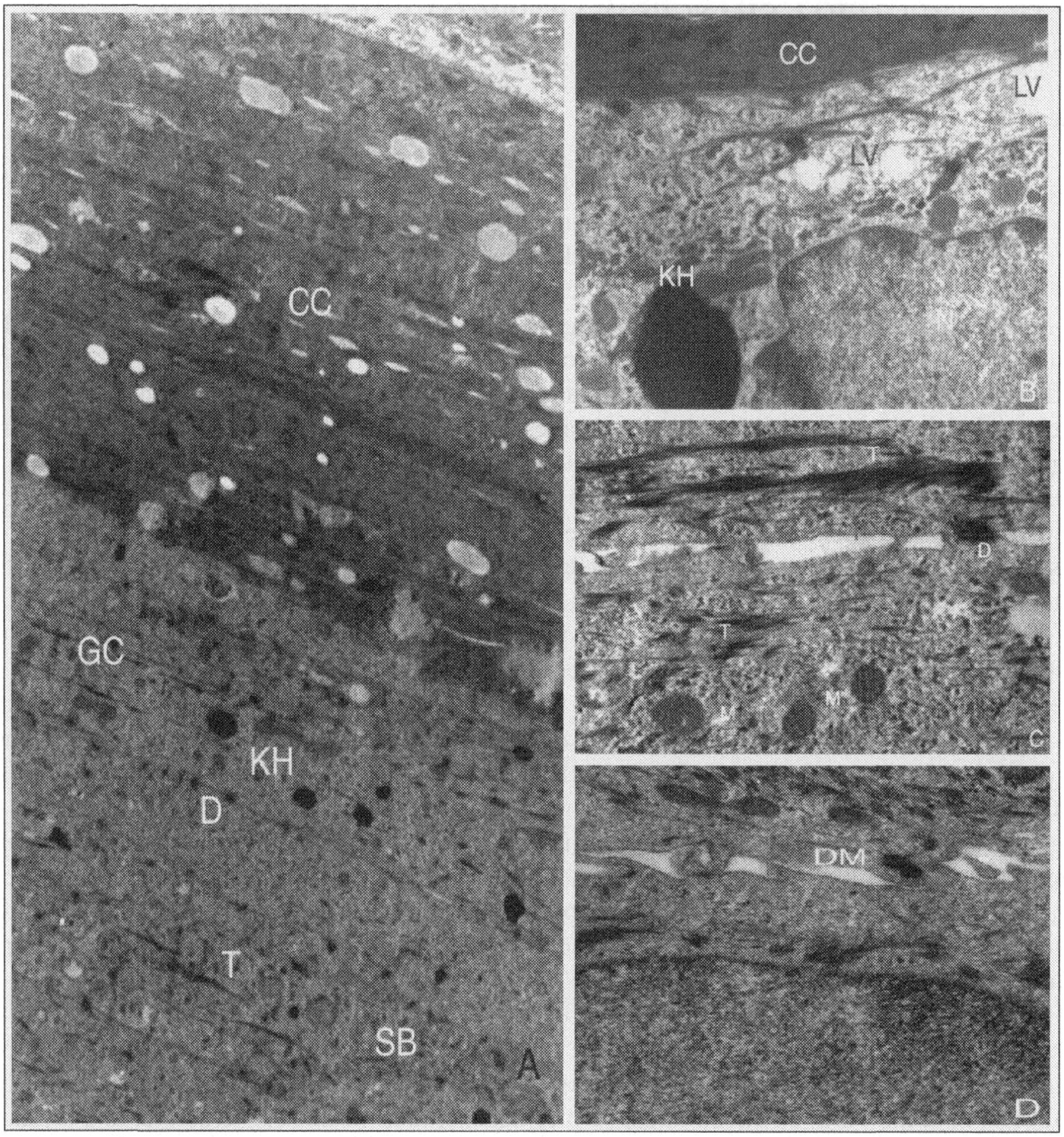

Figure 3.7 Electron micrograph of cultured skin, topically exposed to As(V) for 6 (A) and 24 (B, C, D) hr. Human keratinocytes were grown submerged for 5 days and lifted for 12 days.

A: Cross-section through the suprabasal (SB), granular (GC) and cornified (CC) layers. Cultures were exposed to 100 µg/L of As(V) for 6 hr. Keratohyalin-like granules (KH), desmosomes (D), tonofilaments (T). Magnification ×4600. No morphological changes were noticed.

B–D: Cultures exposed to 100 µg/L of As(V) for 24 hr; Magnification ×19000.

B: Compact cornified cell layers (CC), empty lipid vacuoles (LV), keratohyalin-like granules (KH), nucleus (N).

C: Note thick tonofilament bundles (T). Some of them lost their fibrillar appearance. Normal mitochondria (M) and desmosomes (D).

D: Mitochondria at different stages of deterioration (DM).

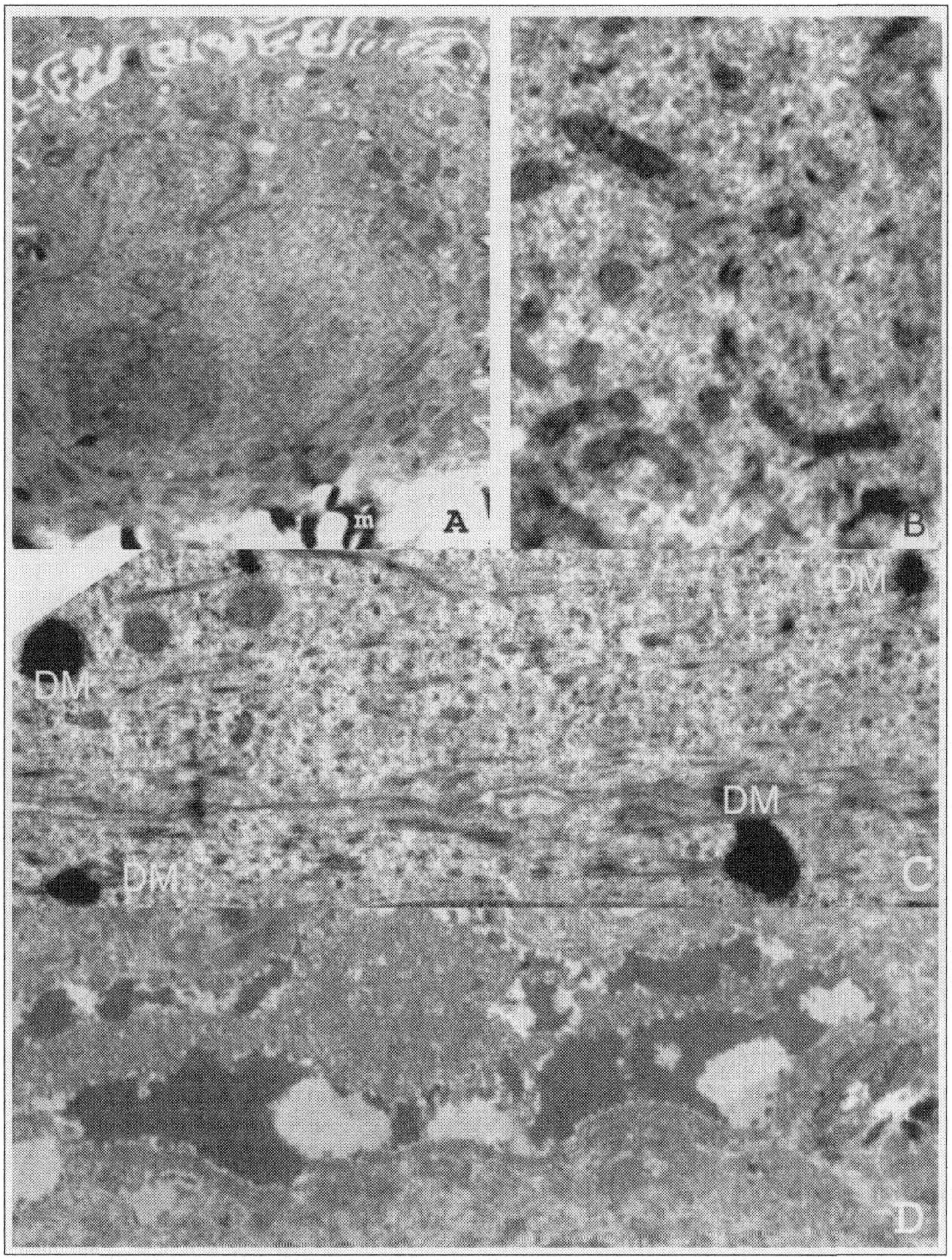

Figure 3.8 Electron micrograph of cultured skin, topically exposed to As(III) for 6 hr. Human keratinocytes were grown submerged for 5 days and lifted for 12 days.
A–B: Cultures, exposed to 10 µg/L of As(III). Magnification ×8706; Note large intracellular spaces (A) and enhanced amount of mitochondria (B).
C: Cultures, exposed to 100 µg/L of As(III). Magnification ×8706; Degenerated mitochondria (DM).
D: Cultures, exposed to 500 µg/L of As(III). Apoptotic nuclei in suprabasal layers. Magnification ×13500.

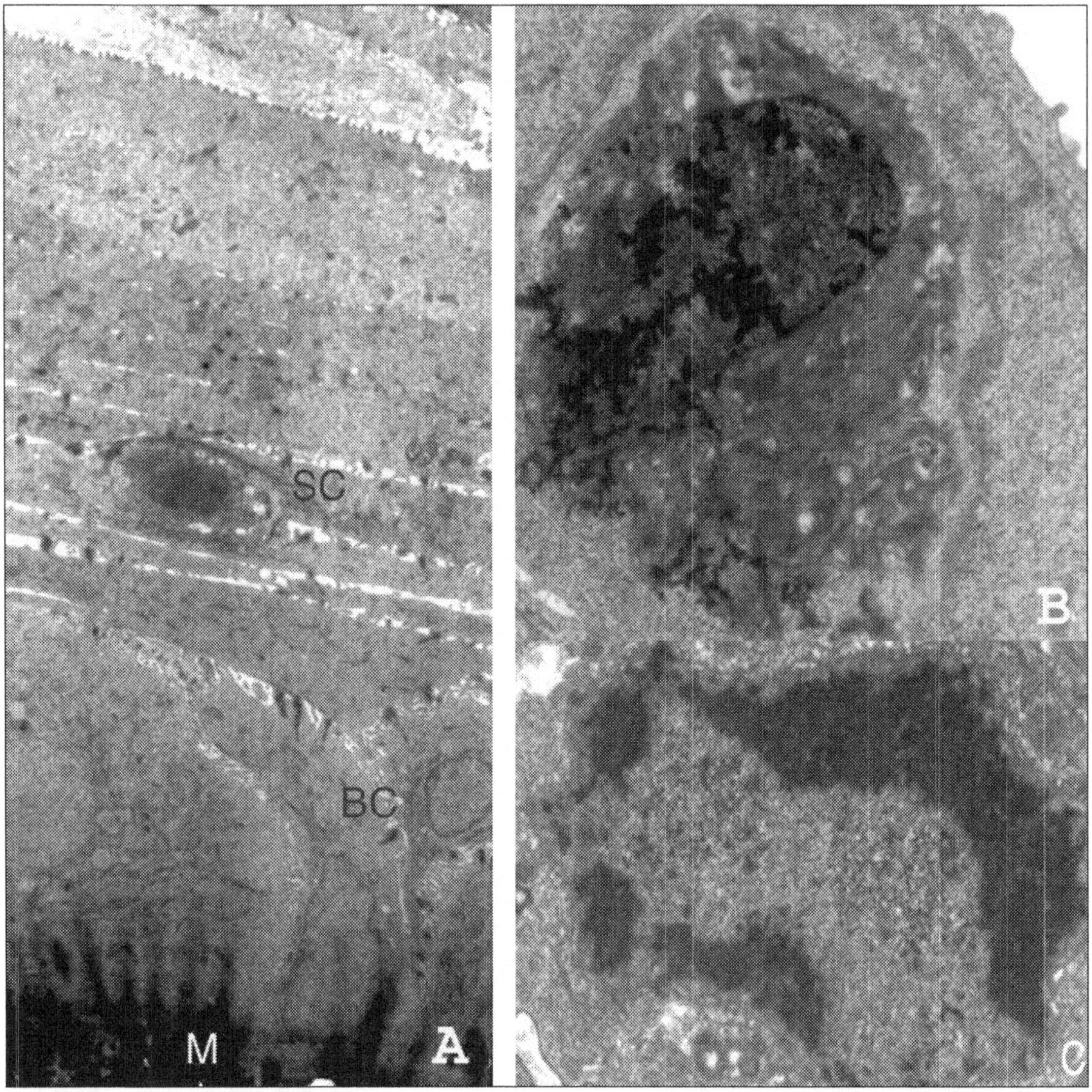

Figure 3.9 Electron micrograph of cultured skin, topically exposed to As(III) for 24 hr. Human keratinocytes were grown submerged for 5 days and lifted for 12 days.
A: Cultures exposed to 100 µg/L of As(III); vacuoles in early spinous cells. Deterioration of nuclear structure of basal cells, absence of keratohyalin-like granules, loss of cornified cells. Magnification ×4600.
B–C: Cultures, exposed to 500 µg/L of As(III). Note nuclear degeneration (B); magnification ×10500. Irregular clumping of nuclear chromatin (C); magnification ×13500.

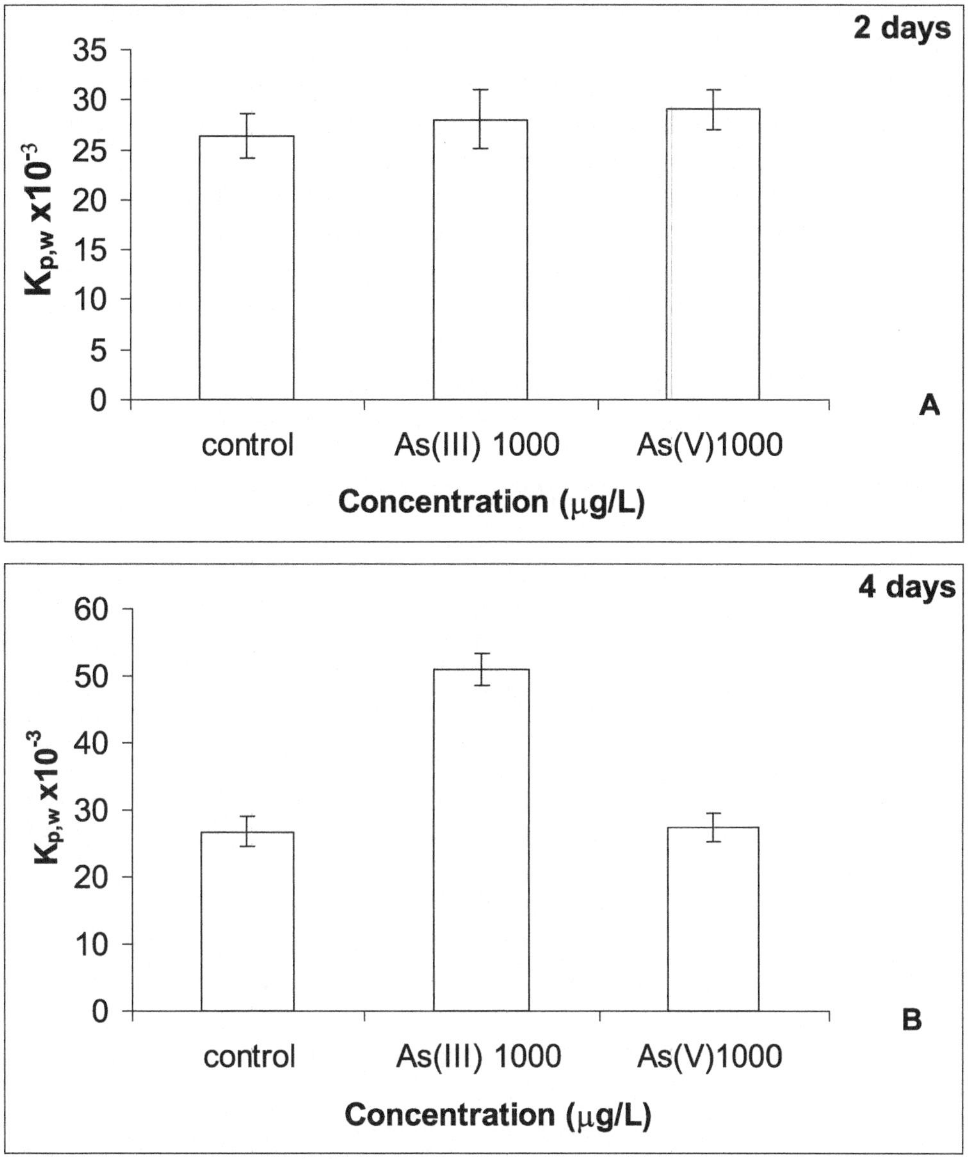

Figure 3.10 Comparative effect of topical exposure to 1000 μg/L of As(III) and As(V) for 2 and 4 days on formation of water permeation barrier in artificial skin. Each point represents the mean of three separate experiments.

PERMEATION AND UPTAKE OF WATER IN ARTIFICIAL SKIN

Artificial skin was tested for the development of water barrier by measuring the $K_{p,w}$ (permeation coefficient for water). We determined the rate of permeation of tritiated water through artificial skin in order to evaluate the integrity of the water barrier of every preparation. In unexposed artificial skin, $K_{p,w}$ was found to be 25×10^{-3} cm/h (standard deviation =3.6; n=20). No difference in $K_{p,w}$ was observed in the presence of 1000 μg/L of As(V) or As(III) after first 24 to 48 hours of topical exposure (Figure 3.10A). After 96 hr of topical exposure, the $K_{p,w}$ for solution of 1000 μg/L As(III) was increased almost 2-fold whereas the value for As(V) solution over the same period of time remained unchanged (Figure 3.10B).

There should, of course, be certain differences between the natural skin and any culture system. For instance, epidermal desquamation is unlikely to occur in the culture as it does *in vivo,* and there is no active vascular system in a culture as there is *in vivo.* Artificial skin, produced by primary keratinocytes, isolated from the newborn rats, resulted in 10% of the cultures with $K_{p,w}$ of 3.2, 30% of the cultures with $K_{p,w}$ of 3.8 and 79% of the cultures with $K_{p,w}$ of $5.5 \times 10^{-3} \pm 1.6 \times 10^{-3}$. These values are 2-3 times higher than those found in newborn rat epidermis (Pu et al., 1995). Artificial skin, produced from adult human skin is 10-fold more permeable to tritiated water than intact human skin. Stability of the water barrier of the epidermis of newborn rats is constant during first 5 days and declines rapidly after that. The available information do not point to a likely difference in the $K_{p,w}$ of human skins for the age group used in the study.

UPTAKE OF AS(III) AND AS(V) BY THE ARTIFICIAL SKIN

During the 96-hr exposure, permeation of As(III) increased almost 2-fold whereas permeation of As(V) over the same period of time remained unchanged. These observations were consistent with the morphological changes observed by phase contrast and in paraffin cross-sections of lifted cultures. Experiments showed that the skin cultures retained most of the applied arsenic when the concentration was less than 500 μg/L (Table 3.2). The rate of retention of the applied arsenic (<500 μg/L) was estimated to be 6-9% per hour (Table 3.1). Changes in concentrations of As(III) and As(V) at hourly intervals with the starting charges of 1000 μg/L are compared in Table 3.1. Most of the arsenic getting into the skin cultures during the experiment was retained during the first hour (As(V)) or second hour (As(III)).

For As(III) and As(V) concentrations of 10, 100, 250 and 500 μg/L, the absorption rate equaled exposure dose; no arsenic permeates through the skin was detected in the collection solution. In other words, all the applied arsenic was absorbed by the cultures. For solutions containing 1000 μg/L As(V), only ~30% of the applied arsenic was absorbed by the culture (Figure 3.11). Experiments with As(III) showed that a higher percentage absorbed per hour for all concentrations as compared to As(V). Artificial skin was first pre-exposed to arsenic from the bottom, rather than applied topically, for 10 days. Various levels of arsenic were then applied topically to the pre-exposed cultures, and the permeation rates were determined (Tables 3.1 and 3.2). Results show a decrease in the percentage of As absorbed per hour for all concentrations of As(III) and As(V). The concentrations of arsenic in collection solutions were detectable for membranes exposed to 100 and 500 μg/L of either As(III) or As(V). This suggests that when cultures were exposed to arsenic from the bottom, damage occurred which impaired the ability of the cultures to retain arsenic. However, when membranes were pre-exposed to As(III) from the bottom for only 24 hours, arsenic could not be detected in the effluent for solutions containing 100, 500 and 1000 μg/L. When exposed culture membranes were digested with concentrated nitric acid, most of the retained arsenic was recovered. This showed that after 24 hours, the culture cells were still able to accumulate arsenic and damage might be the result of "long-term" sustained exposure to "low doses" of arsenic. When the Nylon membranes used to retain the cultures were leached with acid and analyzed, no arsenic was detected.

Table 3.1
Absorption of As(III) by skin cultures

Conc.		Volume collected per hour (mL)									
Channel	Blank	Hour 1	Hour 2	Hour 3	Hour 4	Hour 5	Hour 6	Volume left (μl)	Quantity applied (μg)	% absorbed per hour	Measured (expected) conc.*
2: 100 μg/L	1	0.5	1.9	1.7	2.1	2.1	1.1	220	0.018	7.5	BDL (1.8 μg/L)
250 μg/L											
2	0.9	2.2	1.6	2.1	2.3	2.2	2.1	215	0.046	9.25	BDL (3.7 μg/L)
500 μg/L											
2	0.5	2.3	1.8	2.2	2.4	2.2	2.2	246	0.077	6.4	BDL (5.9 μg/L)
3	1	0.8	2.5	2.1	2.5	2.3	1.7	250	0.075	7.5	BDL (6.3 μg/L)
1000 μg/L	2	2.5	2.1	2.1	2.7	2.4	1.8	192	0.21	8.7	Hr1: BDL (15.4 μg/L) Hr2: 2
As(V)	2.4	2	2	2	2.5	2	1.5	38	0.36	18.1	Hr1: 3 (30 μg/L) Hr6: 12

* BDL= Below instrumental detection limit (<2 μg/L); Calculated (expected) concentration assumes constant topical application of arsenic to the culture.

ARSENATE REDUCTION BY AMINO ACIDS

The mechanism of arsenate reduction in skin cells is unclear. Although a number of enzymatic pathways have been proposed, reduction by non-enzymatic processes has received little attention. This study shows that in addition to glutathione and cysteine, which has been reported previously (NAS, 1999), non-enzymatic reduction of arsenic by homocysteine is also possible. Like glutathione ($C_{10}H_{17}N_3O_6S$) and cysteine ($C_3H_7NO_2S$), homocysteine ($C_4H_9NO_2S$) also contains sulfur. We find that not all amino acids containing sulfur would reduce arsenic chemically. Methionine ($C_5H_{11}NO_2S$) did not reduce As(V), for instance.

Experiments with an amino acid:As(V) ratio of 2:1 yielded results that indicated the slow reduction of As(V) by homocysteine. As(V) was reduced by ~17% (2.68%/day) in the presence of homocysteine after 7 days. The reduction rate is shown in Figure 3.12, and described by the equation $y = -0.0004x^2 + 0.1664x + 0.0739$, $R^2 = 0.9987$. With glutamine, methionine, and alanine, no reduction of As(V) was observed (Figure 3.12). Analysis of samples from controls each day did not detect any As(III). Temperature was maintained at 37°C and pH of the solutions was adjusted to remain in the range of 2.7 to 2.8 through the duration of the experiments thereby eliminating temperature and pH as factors in arsenate reduction. Increasing the homocysteine/As(V) ratio greatly increases the reduction rate (Figures 3.13 and 3.14). These results show that reduction of As(V) by proteins is possible at physiological temperature (though not physiological pH), suggesting that arsenate can be converted to arsenite non-enzymatically by reduced species with the skin cells.

Table 3.2
Percutaneous absorption of As(III) and As(V) by artificial human skins

As conc. (μg/L)	Applied dose absorbed (%/hr)*	Calculated K_p (cm/hr)	Arsenic absorption rate (μg/(cm^2/hr))†	Estimated dermal exposure (μg/(kg/15 min))‡	
				Child (8 yrs old)	Adult
As(V)					
10	2.5	3.1×10^{-5}	0.31×10^{-3}	0.002	0.002
100	1.3	1.62×10^{-5}	1.62×10^{-3}	0.012	0.009
250	4.7	5.88×10^{-5}	14.7×10^{-3}	0.106	0.086
500	5.4	6.77×10^{-5}	33.8×10^{-3}	0.239	0.193
1000	7.5	$38.5\text{x}10^{-5}$	385×10^{-3}	0.66	0.53
		4.3×10^{-3} (Mean)§			
As(III)					
100	7.5	9.38×10^{-5}	9.38×10^{-3}	0.066	0.054
250	9.2	9.64×10^{-5}	24.1×10^{-3}	0.204	0.164
500	6.4	7.94×10^{-5}	39.5×10^{-3}	0.282	0.228
1000	7.0	13.4×10^{-5}	130×10^{-3}	0.62	0.49
		10.1×10^{-3} (Mean)			

* Derived from present study
† Assuming thickness of water (wet) film on the body is 100 μL
‡ Assuming average surface area of an adult is 20,000 cm^2 and average weight of 70 kg. Assumed surface area for 8-year-old child is 10,700 cm^2 and average weight is 28.2 kg (USEPA, 1996)
§ Excluding the data for the 1000 μg/L dose

CYTOTOXICITY EVALUATION IN SUBMERGED CULTURES

LDH Assay

For these experiments, 24-hr old human epidermal keratinocytes were exposed to 10, 100, 250, 500 and 1000 μg/L of As(V) or As(III). In general, LDH leakage in cultures, exposed to As(V) was lower than that from cultures exposed to As(III) when compared at the same exposure dose, and there was a tendency for the observed effects to be similar after longer periods of exposure. The LDH release was increased above control as early as 30 min of exposure to 500 μg/L of As(V). At 60 min exposure, LDH leakage was increased beginning with exposure to 100 μg/L and there was a marked increase at 500 and 1000 μg/L of As(V). At 2 hr exposure, the submerged cultures uniformly showed increased LDH release by 20-30% regardless of the As(V) exposure level

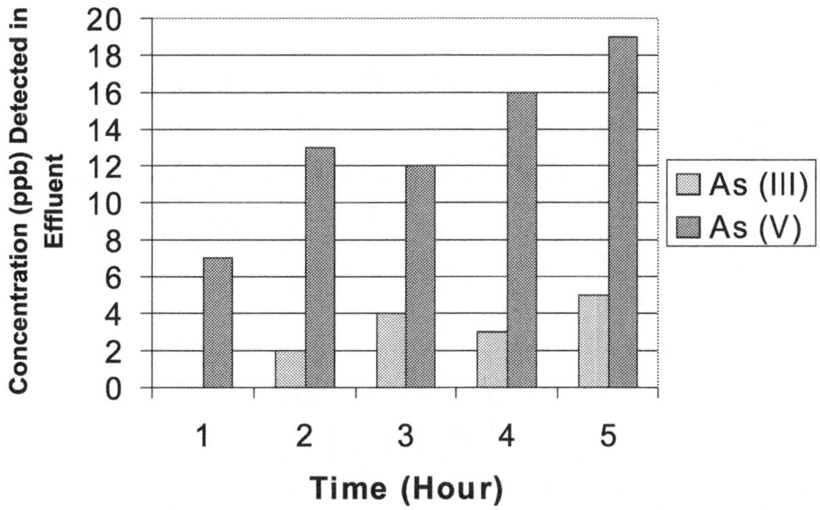

Figure 3.11 Permeability studies using 1000 µg/L As show that ~70% As(V) passes through the culture and ~30% is absorbed by the culture, and ~10% As(III) passes through the culture and ~90% is absorbed by the culture

	% Reduction			
Time (hours)	Homocysteine	Glutamine	Methionine	Alanine
45	7.02	BDL	BDL	BDL
72	9.80	BDL	BDL	BDL
96	12.4	BDL	BDL	BDL
142	15.1	BDL	BDL	BDL
169	16.7	BDL	BDL	BDL

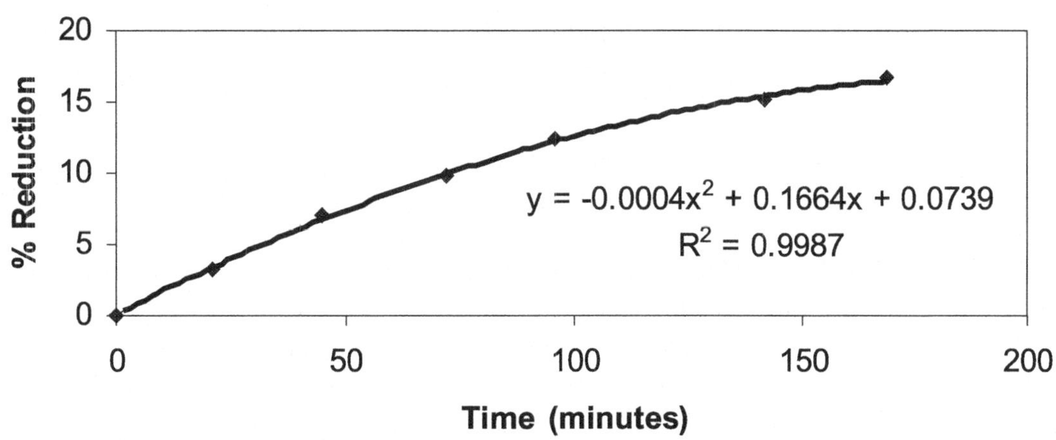

Figure 3.12 Arsenate reduction by homocysteine, glutamine, methianine and alanine over 7 days; note that the homocysteine/As(V) ratio was 2/1. Data plotted are tabulated above the figure along with results of experiments using glutamine, methionine and alanine. BDL = below detection limit.

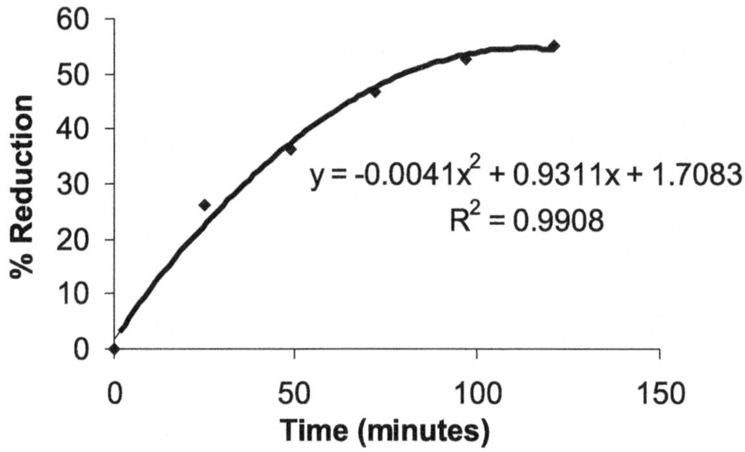

Figure 3.13 The effect of 10× increased homocysteine concentration on As(V) reduction

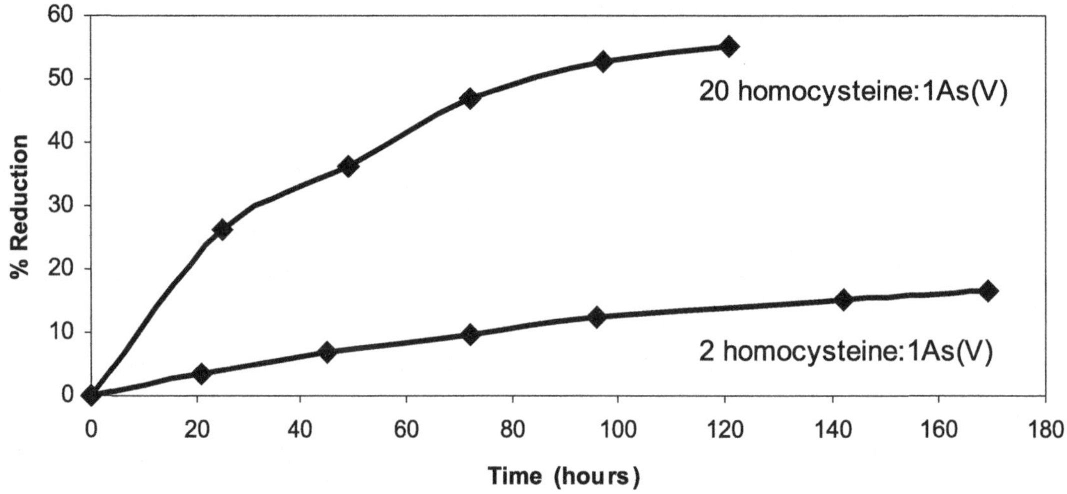

Figure 3.14 The effect of 20× increased homocysteine concentration on As(V) reduction

(Figure 3.15). Longer exposure to different doses of As(V) showed a lot of scatter in the results, with only the highest doses of 500 and 1000 μg/L displaying statistically significant differences (Figure 3.16).

The LDH release from HK exposed to various concentrations of As(III) revealed a dose-dependent pattern of response. With longer exposures, the release was found to occur at lower concentrations of As(III); approximately (20%) LDH release was observed after 30-min exposure to 100 μg/L (Figure 3.17A), whereas similar effect was noted with 1 μg/L after 120 min exposure (Figure 3.17B). As(III) concentrations of 0.01-0.05 μg/L applied for up to 120 min did not produce statistically significant changes (relative to controls) in submerged 24-hr old HK cultures, but concentrations as low as 0.5-1.0 μg/L As(III) resulted in a significant increase in LDH, which remained unchanged over the 120 min duration of exposure (Figure 3.17B). These tests show that

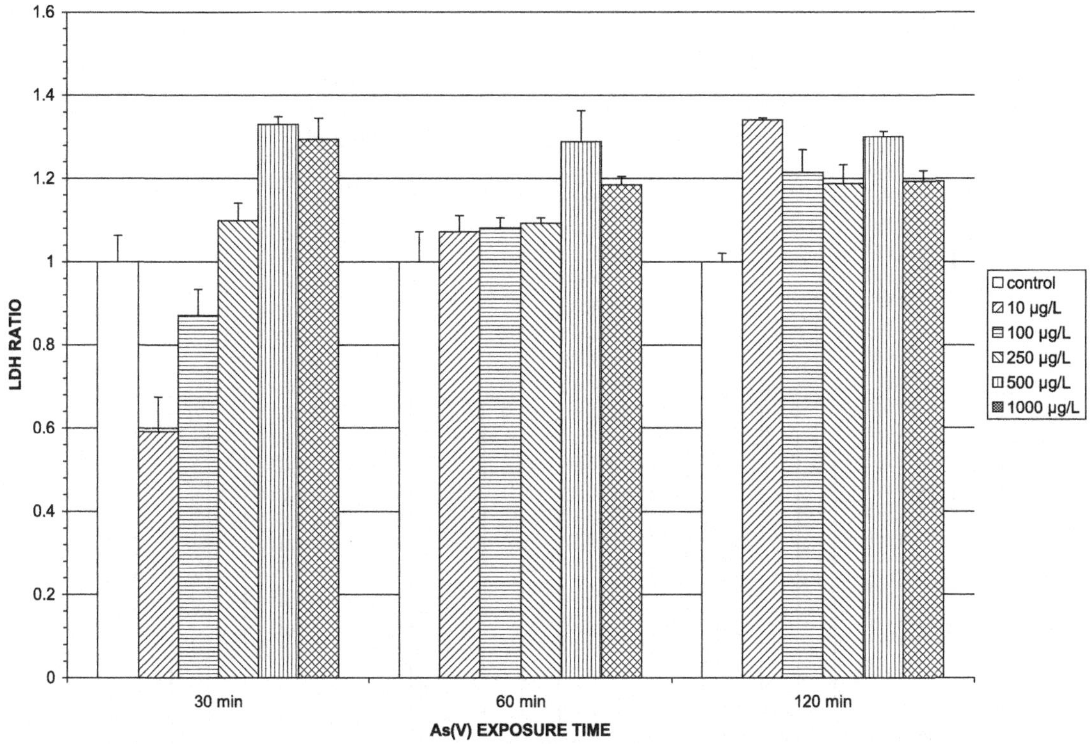

Figure 3.15 Analysis of cytotoxicity measured by LDH release from submerged cultures of human keratinocytes exposed to As(V) for 30, 60 and 120 minutes. Results are expressed as ratio of untreated control. Each point represents the mean of three separate experiments.

the integrity of cell membranes and mitochondrial function can be impaired by exposure to very low doses (<1.0 µg/L) of As(III).

MTS Cell Viability Assay

MTS assay of cells exposed to As(V) for 48 hr and up to 1week did not reveal any statistically significant changes over the 10-1000 µg/L concentration range. By contrast, exposure to As(III) produced a notable decrease in cell viability, ranging from about 20% of control at 10 µg/L to about 40% below level of control at 1000 µg/L exposure for 48 hr and 80% below level for control at 1000 µg/L after exposure for 1 week (Figure 3.18 A, B).

CYTOTOXICITY EVALUATION IN ARTIFICIAL SKIN

Exposure of artificial skin to 500 µg/L As(V) resulted in LDH leakage that was about 2-fold higher than that of unexposed control cultures after one hour. A 5-fold increase in the LDH release was seen in cultures dosed with 1000 µg/L As(V) within the first hour of exposure (results not shown). The higher toxicity of As(III) compared with that of As(V) could be seen in an

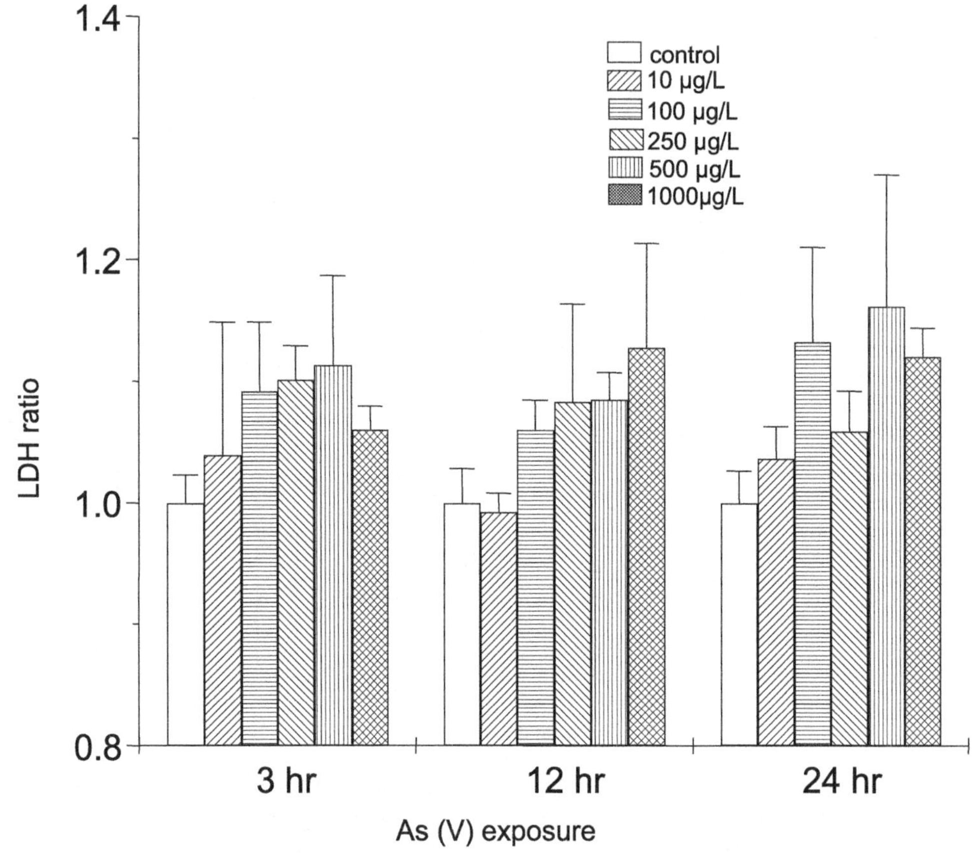

Figure 3.16 Analysis of cytotoxicity measured by LDH release from submerged cultures of human keratinocytes exposed to As(V) for 3, 12 and 24 hours. Results are expressed as ratio of untreated control. Each point represents the mean of three separate experiments.

increase of LDH release during 1-hr exposure to 250 µg/L of As(III) (Figure 3.19), whereas the As(V) concentration had to be no less than 500 µg/L to induce the same effect. In general, changes in LDH release produced by exposures to either As(V) or As(III) occurred within the first 2 hours.

Incorporation Studies in Artificial Skin

These experiments were performed in artificial skin with samples removed at set times (0.5, 1, 2, 3 and 24-hr) to measure the level of incorporation of ^{3}H-thymidine (DNA synthesis), ^{3}H-uridine (RNA synthesis) and ^{14}C-leucine (protein synthesis) as affected by exposure to 10–500 µg/L of As(V) or As(III). In assessing any effects related to protein levels in exposed submerged cultures of HK, it should be noted that arsenic exposure in some cases affected the amount of extractable protein apparently due to both degradation processes and leakage of proteins from damaged cells.

In general, the exposure of submerged HK to As(III) affected the quantity of the total extractable protein. The reductions in the extractable protein were 10-25% at 10 µg/L and 45% at 100 µg/L after 3 days (Figure 3.20). Longer exposure of HK to As(III) of up to 6 days

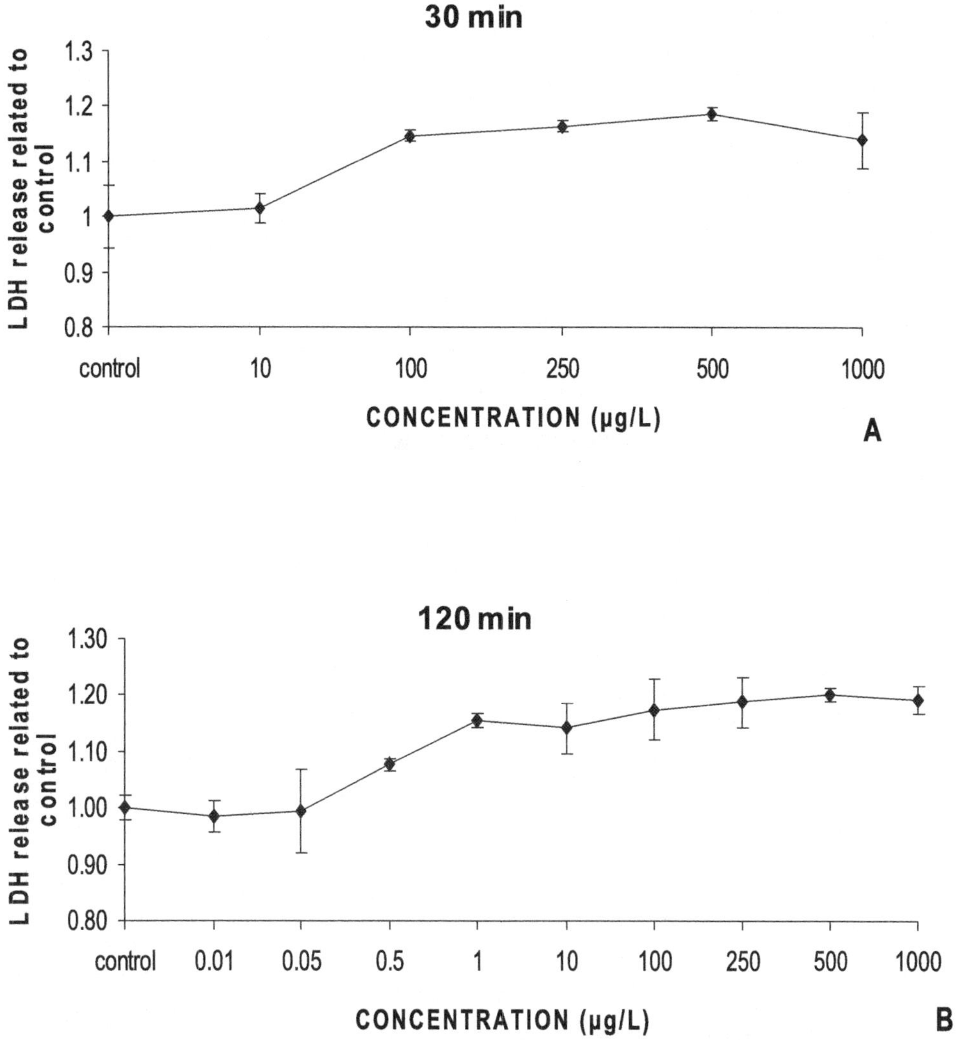

Figure 3.17 Analysis of cytotoxicity measured by LDH release from submerged cultures of human keratinocytes exposed to As(III) for 30 min (A) and 120 minutes (B). Results are expressed as ratio of untreated control. Each point represents the mean of three separate experiments.

did not induce additional loss in protein at 10 and 100 µg/L. However, with prolonged exposure (3–6 days) to 500 and 1000 µg/L As(III), the noticeable loss in extractable protein was associated with morphological changes in HK and the loss of structural integrity of cells. By contrast, HK cultures exposed to As(V) initially developed protein loss (25–27% at 10 µg/L) that was followed by gradual increase in protein concentrations at higher doses of As(V) (14% at 1000 µg/L for 1 hour and 53% after 3 hours) (figure not shown).

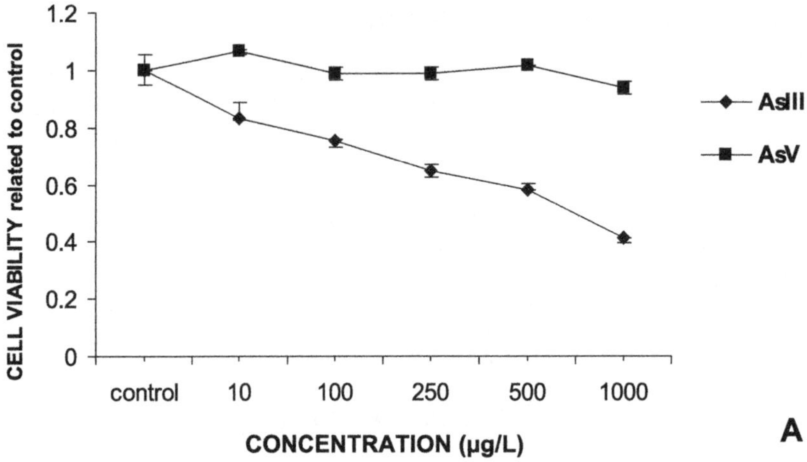

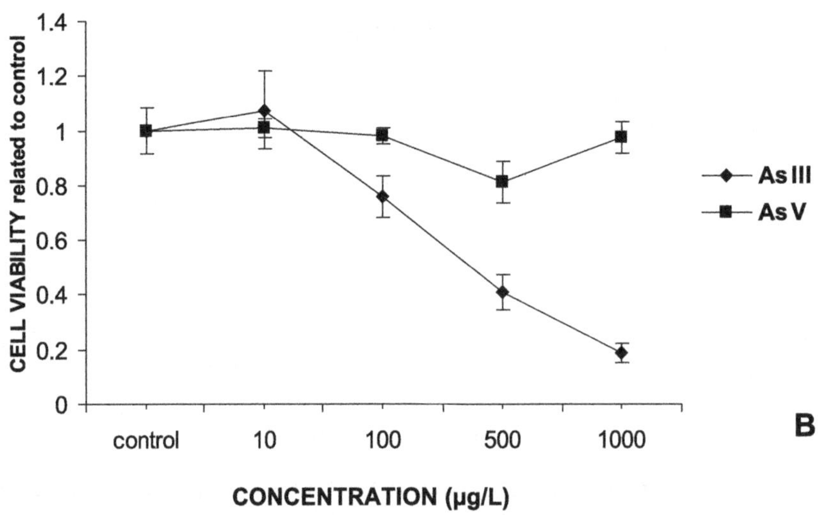

Figure 3.18 Analysis of viability measured by MTS assay. Confluent cultures of human keratinocytes were exposed to As(III) and As(V) for 48 hr (A) and 7 days (B). Results are expressed as ratio of untreated control. Each point represents the mean of three separate experiments.

DNA Synthesis

Exposure of artificial skin to any level of As(V) used in the study revealed about 50% reduction in DNA synthesis within the first 30 min (Figure 3.21A) and a stimulation of ^{3}H-thymidine incorporation at 2 hr exposure to 100 and 500 ug/L of As(V) (Figure 3.21A). In experiments in which the level of ^{3}H-thymidine incorporation was measured following a brief (2-hr) repair period (i.e., after pre-exposure to As(V)), it was noted that the incorporation levels were initially reduced by significant amounts at all concentrations of As(V) (Figure 3.22). After 2-hr repair period, the incorporation practically returned to control levels. After 24-hr repair, there was

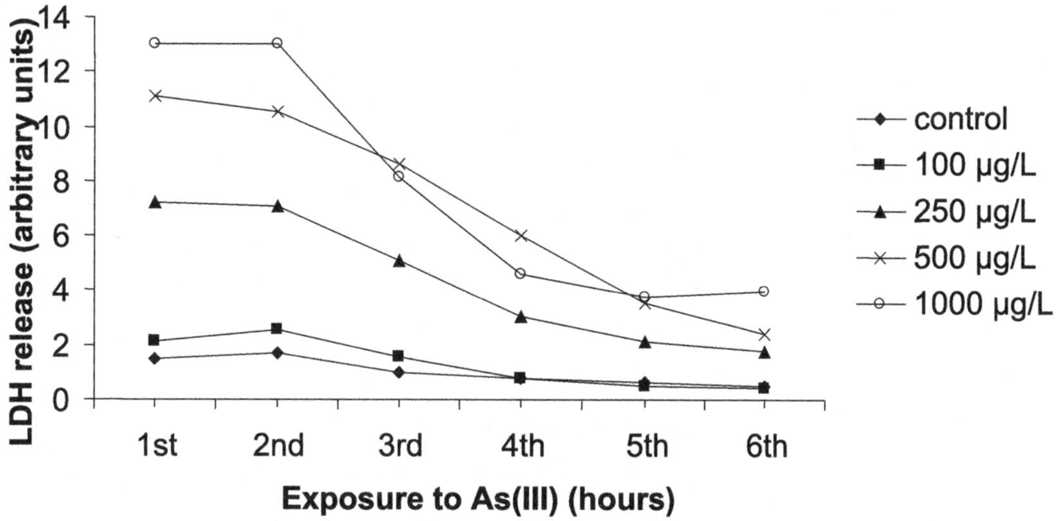

Figure 3.19 Effect of As(III) on the LDH release in the eluate collected during permeability studies. Artificial skin was topically exposed to 100, 250, 500 and 1000 µg/L of As(III) in 0.85% NaCl in the permeation cell. This figure represents the results of a typical experiment where each point is the mean of three independent determinations.

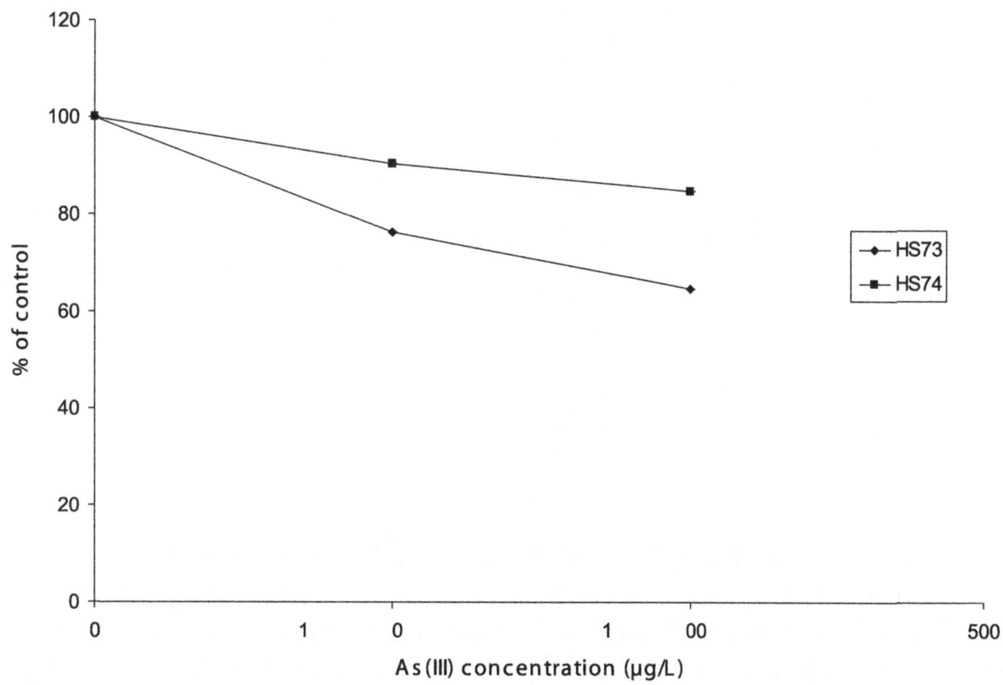

Figure 3.20 Effect of As(III) on the total protein level in exposed submerged cultures. Upper curve (▪) = exposure for 48 hours; lower curve (♦) = exposure for 96 hours. The HS73 and HS74 refer to culture number.

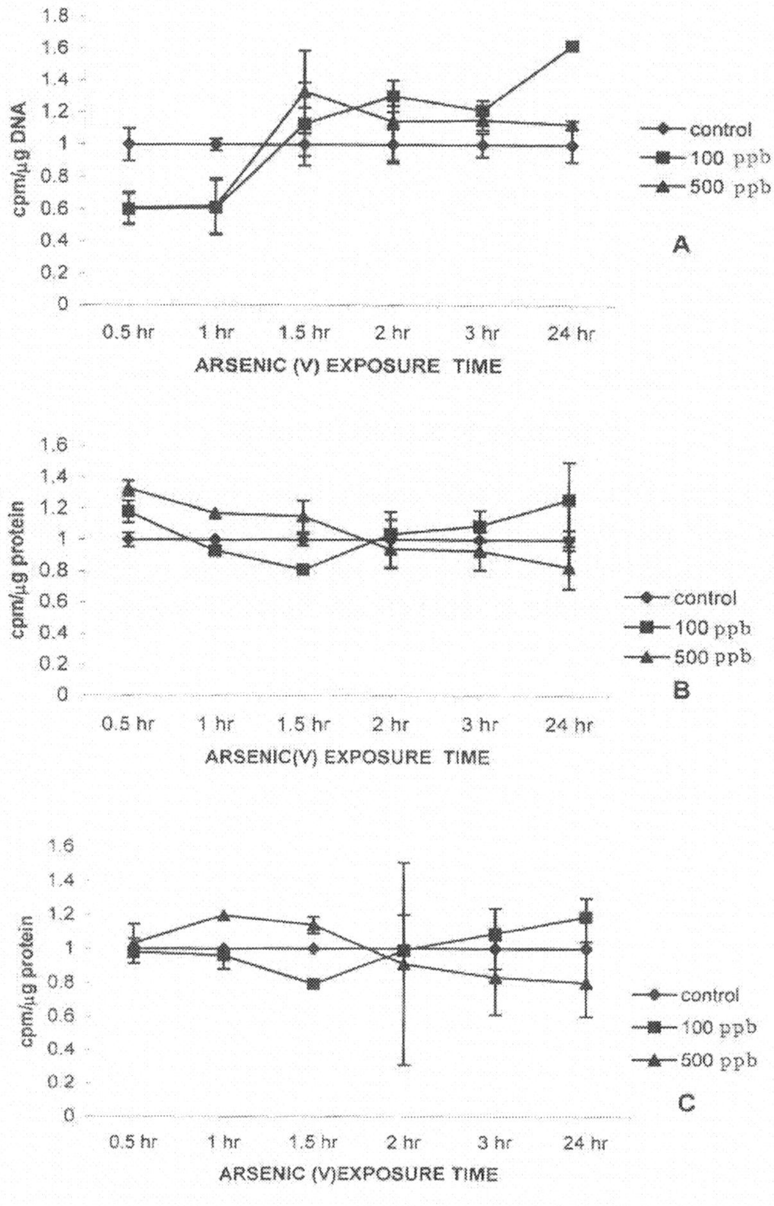

Figure 3.21 Effect of topical application of As(V) on DNA, RNA and protein synthesis in artificial skin expressed as specific radioactivity of respective precursors. After exposure, the cultures were pulse-labeled for 90 min with radioactive precursors at the times indicated, harvested and the TCA acid-insoluble radioactivity was determined. Results are expressed as percent of untreated control. Each point represents the mean of three separate experiments using triplicate cultures.
A: ^{3}H-thymidine incorporation; B: ^{3}H-uridine incorporation; C: ^{14}C-leucine incorporation.

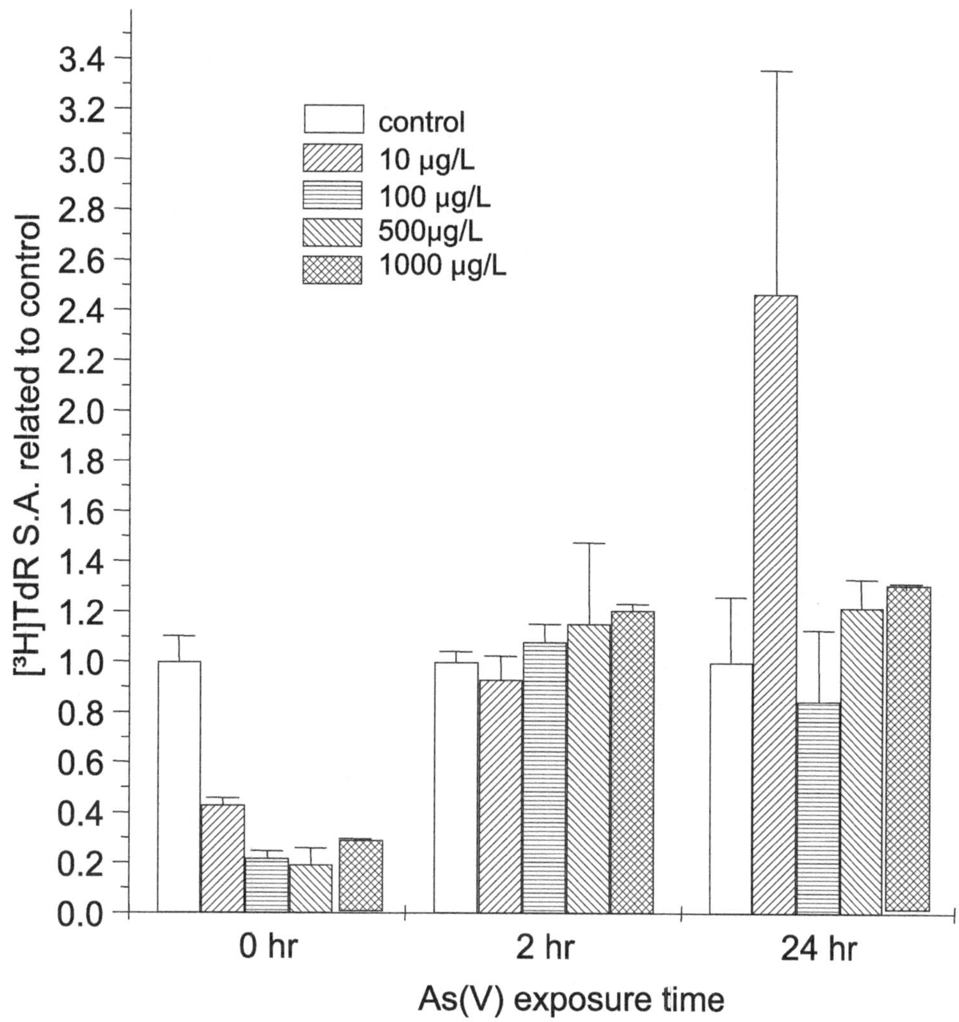

Figure 3.22 Specific radioactivity of ^{3}H-thymidine incorporation in artificial skin following periods of repair after a 2-hr pre-exposure to As(V)

a notable increase in ^{3}H-thymidine incorporation in cells pre-exposed to 10 μg/L As(V), while at all other As(V) concentrations, the degree of incorporation was similar to that of control. These were interesting observations deserving further investigation.

Exposure to 10 and 500 μg/L As(III) resulted in increased ^{3}H-thymidine incorporation in cultures within the first two hours (Figure 3.23). The incorporation gradually increased up to the third hour of exposure followed by a leveling off that lasted for as long as 24-hr of exposure. Decreased ^{3}H-thymidine incorporation was produced in 24-hr exposure to 500 μg/L As(III).

The strong effect of As(III) on the biosynthetic activity of keratinocytes was demonstrated in ^{3}H-thymidine incorporation experiments when As(III) was applied from underneath the cultures to simulate *in vivo* toxicity of skin from the bloodstream. Cells were exposed to As(III) for 48 hours from underneath (the arsenic was contained in the growth medium). These experiments

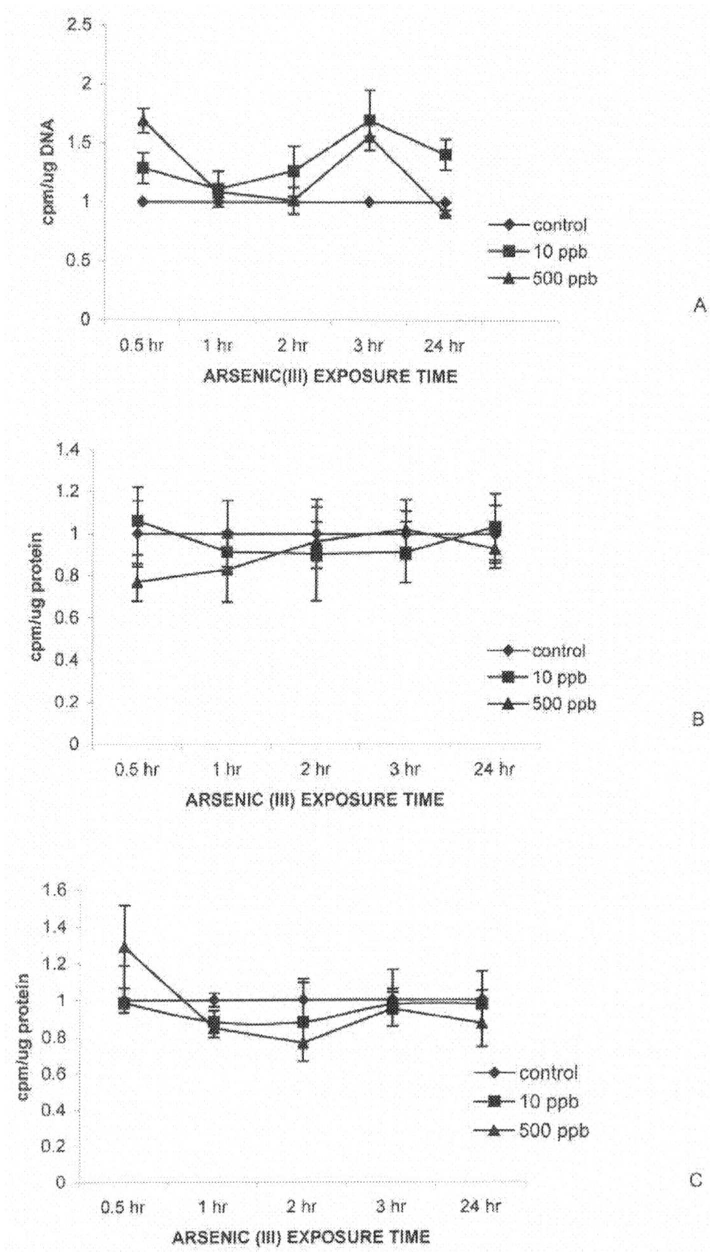

Figure 3.23 Effect of topical application of As(III) on DNA, RNA and protein synthesis in artificial skin expressed as specific radioactivity of respective precursors. After exposure, the cultures were pulse-labeled for 90 min with radioactive precursors at the times indicated, harvested and the TCA acid-insoluble radioactivity was determined. Results are expressed as percent of untreated control. Each point represents the mean of three separate experiments using triplicate cultures.
A: ^{3}H-thymidine incorporation; B: ^{3}H-uridine incorporation; C: ^{14}C-leucine incorporation.

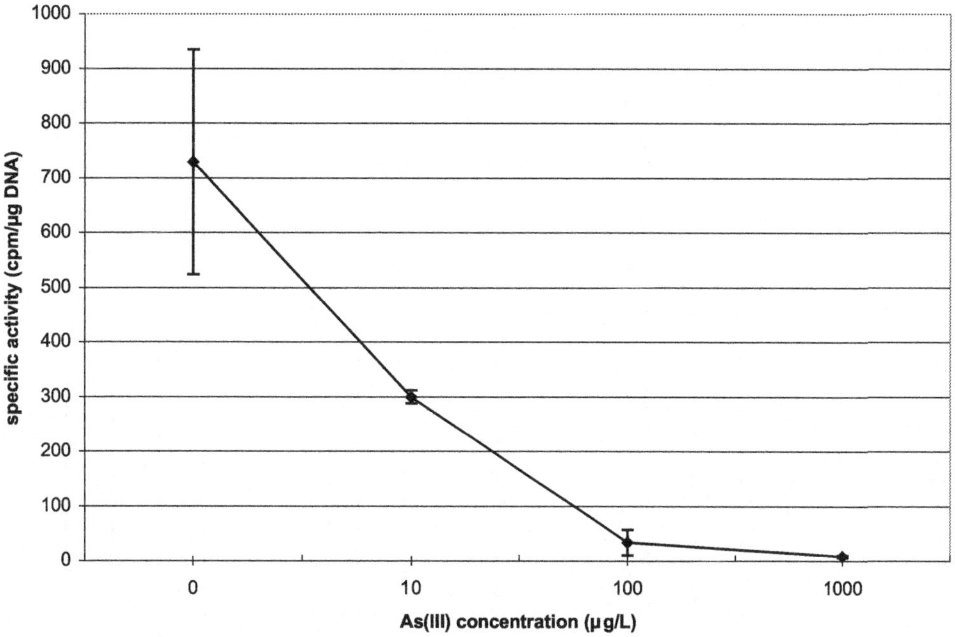

Figure 3.24 Effect of As(III) on the incorporation of ^{3}H-thymidine into DNA in artificial skin exposed to 10, 100 and 1000 µg/L of As(III) from underneath

showed some 69% inhibition of thymidine incorporation at 10 µg/L, 95% at 100 µg/L and 98% at 1000 µg/L (Figure 3.24). The remarkable thing is that As(III) can affect the DNA synthesis in cells at concentrations of 1.0 µg/L. As far as we know, this is one of the lowest effect levels that have been reported for arsenic on human cell culture.

RNA Synthesis

Variability in the pattern of incorporation of labeled ^{3}H-uridine at all concentrations and durations of exposure to As(V) did not show statistically significant differences from that of control cultures (Figure 3.21B). There seemed to be a difference in response by cultures exposed to As(III) in terms of ^{3}H-uridine incorporation (Figure 3.23B). The observed increase in ^{3}H-uridine incorporation, however, was too subdued to yield any insights into the effects of As(III) on RNA synthesis.

^{3}H-uridine incorporation was found to be stimulated by high levels of As(V) (1000 µg/L) immediately after a 2-hr repair period or by a low dose of As(V) (10 µg/L) with the longest repair period (24-hr) (Figure 3.25). The nature of this conflicting stimulation was not clear at the time and presumably was an experimental artifact. The experiment was not repeated.

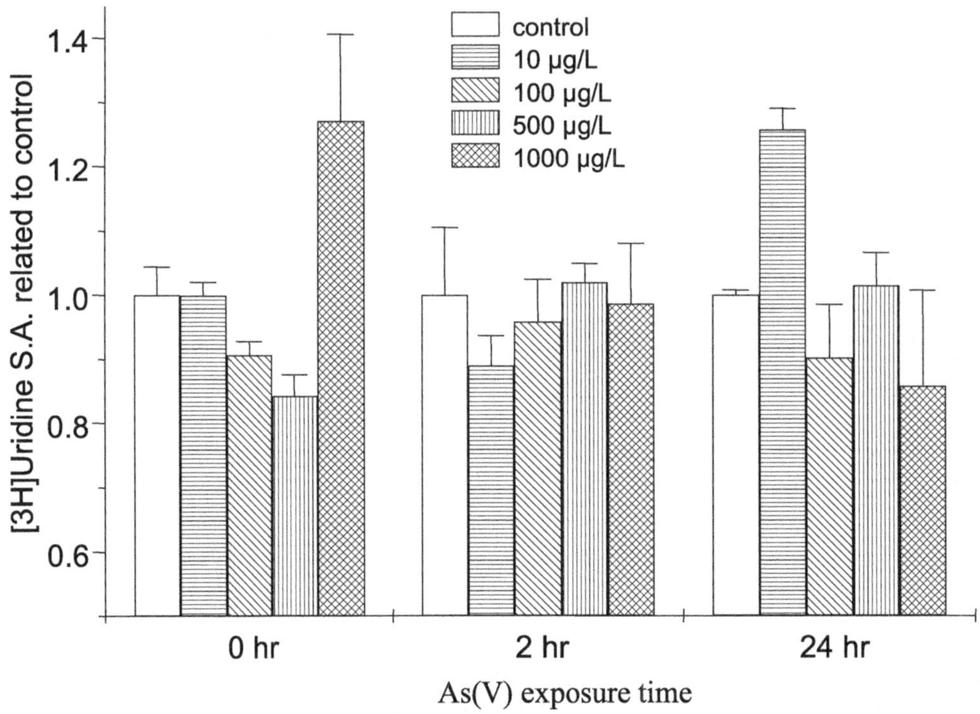

Figure 3.25 Specific radioactivity of ^{3}H-uridine incorporation in artificial skin following periods of repair after a 2-hr exposure to As(V). These experiments were performed in triplicates and the only statistically significant differences observed were at 0 hr, 1000 µg/L and 24 hr, 10 µg/L.

Protein Synthesis

Statistically significant increase in ^{14}C-leucine protein was noted after 1.5-hr exposure to 500 and 100 µg/L As(V) (Figure 3.21C). Similar to DNA synthesis, low dose (10 µg/L) also increased incorporation of ^{14}C-leucine after 2-hr exposure to As(V) (data not shown). Likewise, early stimulation (by up to 50%) of ^{14}C-leucine incorporation into protein could be seen in cultures exposed to 10-500 µg/L As(III) within 30 min. Somewhat longer exposure to all concentrations of As(III) tested (10-1000 µg/L) resulted in decreased protein synthesis (Figure 3.23C).

As noted above, ^{3}H-TdR incorporation was also stimulated (by up 50%), but at different times and doses of exposure to As(III) (Figures 3.22–3.23). Both ^{3}H-TdR and ^{14}C-leucine incorporations returned to control levels by the end of 24-hr exposure for the concentrations of As(III) tested.

Similar to the effect on RNA synthesis, high level (1000 µg/L) of As(V) also increased incorporation of ^{14}C-leucine immediately after 2-hr repair period or by a low dose of 10 µg/L after 24-hr repair period as shown in Figure 3.26. All the incorporation experiments are thus consistent in demonstrating that the effects levels for both As(III) and As(V) are below 10 µg/L, the current national MCL for arsenic in drinking water.

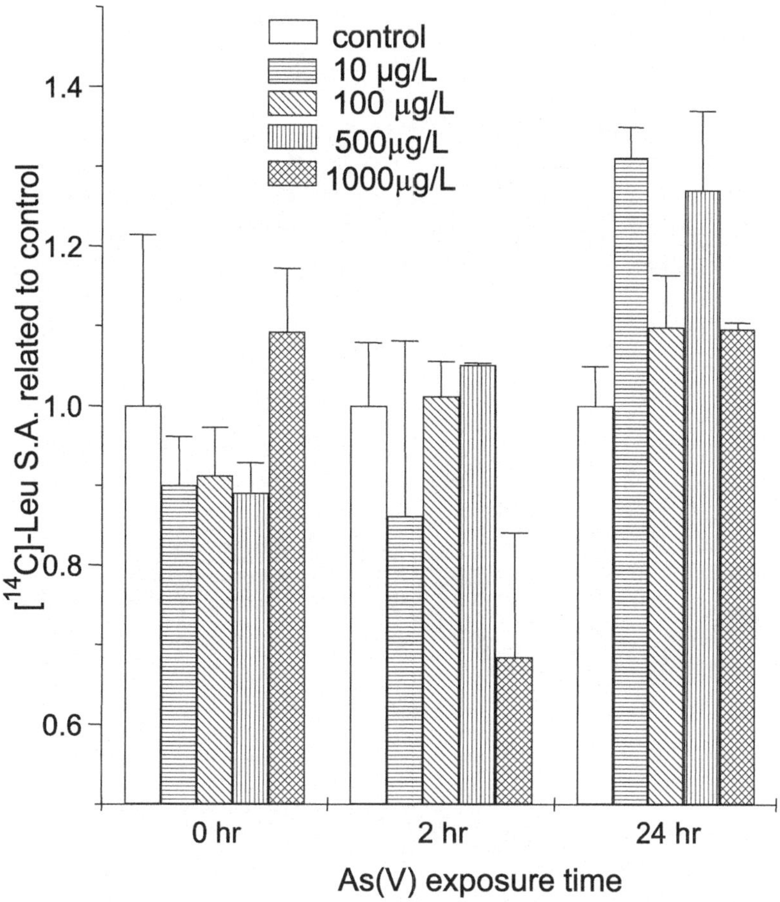

Figure 3.26 Specific radioactivity of ^{14}C-leucine incorporation in artificial skin following periods of repair after a 2-hr exposure to As(V)

EFFECTS OF ARSENIC ON PROGRAMMED CELL DEATH

DNA Fragmentation Assay (TUNEL Assay)

Submerged 24-hr old cultures were exposed for 3 hours to 10–1000 µg/L of As(V). No changes were seen at 10-500 µg/L As(V). Consistent changes, however, were observed at 1000 µg/L As(V) exposure (Figure 3.27).

Artificial skin was topically exposed to 10-1000 µg/L for 1-hr and 6-hr. No apoptotic changes were noticed at 1-hr exposure, whereas significantly increased apoptotic changes could be observed after 6-hr exposures to 100, 500 and 1000 µg/L As(V) (Figure 3.28).

Changes *in situ* DNA fragmentation at the single cell level in lifted cultures of human keratinocytes exposed to various concentrations of As(III) for 1-72 hours are presented in Figure 3.29. The horizontal row of pictures represent a series of tissue sections taken at different time intervals (1 through 72-hr), while the vertical rows represent increasing concentrations of As(III) to which the cells were exposed (10 through 1000 µg/L). The major features can be

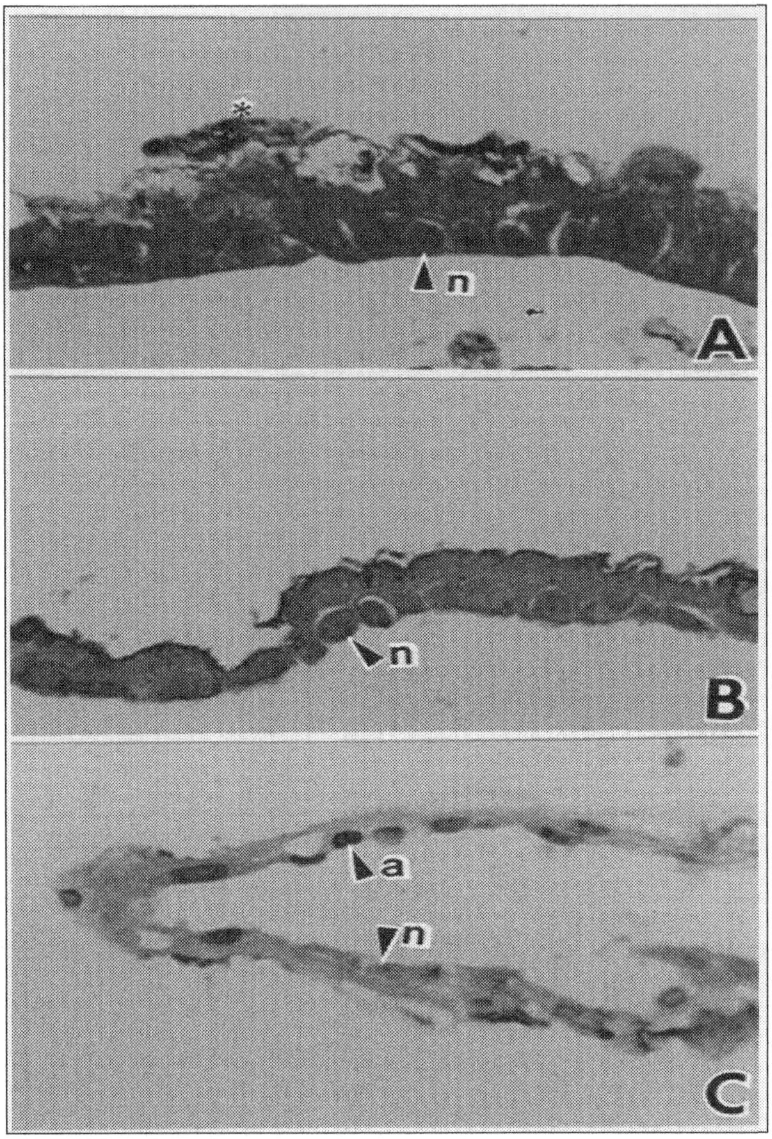

Figure 3.27 Photomicrographs showing dose-dependent changes in submerged cultures of human keratinocytes exposed to As(V) for 3-hr. A: control; B–C: exposure to 500 and 1000 μg/L, respectively. Apoptotic (A) and normal (N) cells are indicated. Staining shows in situ DNA fragmentation at the single cell level.

summarized as follows: the progressive changes in cells exposed to increasing doses of As(III) were manifested in the transition of apoptotic to necrotic changes. With the increase in the time of exposure to a given concentration of As(III), the apoptotic changes not seen at 1-hr of exposure to 10 μg/L became manifested after 6-hr, with progressive deterioration suggestive of necrotic death at 48 and 72-hr exposure. The changes observed at 6-hr exposure to 10 μg/L resemble those found in cells exposed to 100 μg/L for 1-hr. Exposure to higher concentrations (500 and 1000 μg/L) resulted in occasional apoptotic changes within 1-hr of exposure which were replaced by necrotic destruction of cells after 6-hr exposure to these doses. At the higher exposure doses (500 and

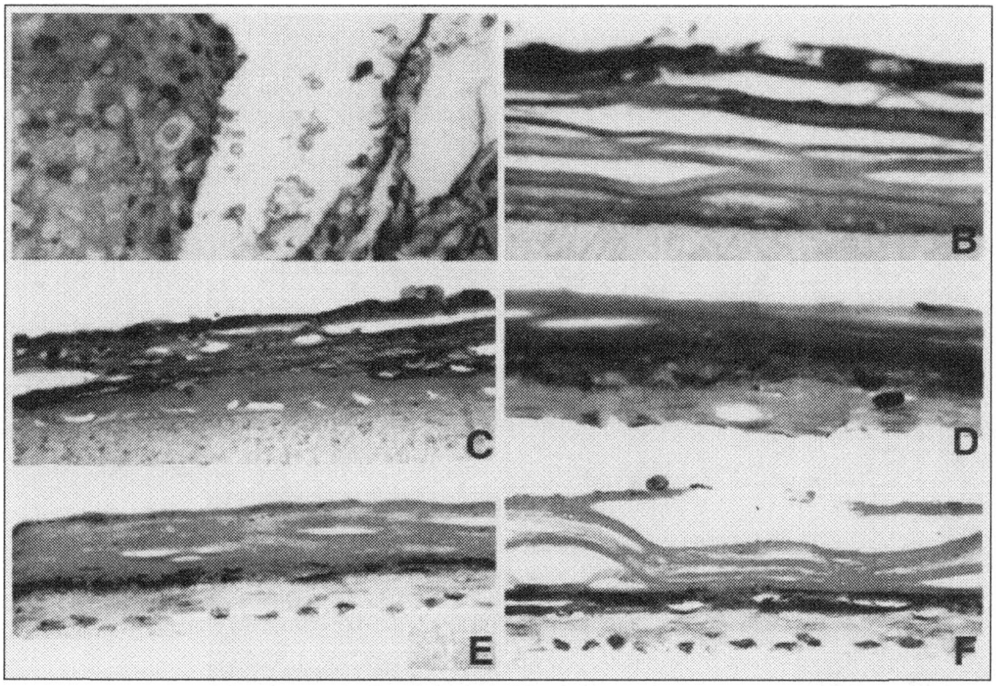

Figure 3.28 Photomicrographs showing dose-dependent changes in lifted cultures of human keratinocytes, topically exposed to As(V) for 1 and 6 hours.
A: Positive control; B: 6-hr exposure to 10 μg/L; C: 1-hr exposure to 100 μg/L; D–F: 6-hr exposure to 100, 500 and 1000 μg/L, respectively. Note dark color of affected nuclei representing apoptotic or necrotic changes. The staining represents specific in situ DNA fragmentation at the single cell level.

1000 μg/L for 72-hr), the cultures were completely destroyed and no structural features could be seen (Figure 3.29, D3-E3).

Apoptosis Evaluation at the Single Cell Level (COMET Assay)

These experiments were conducted in submerged cultures of HF and HK. They could not be performed in lifted cultures since the method called for single cell suspensions that were not obtainable from lifted cultures. We exposed the cells to 10, 100, 500 and 1000 μg/L of As(III) for 30-min, 1, 2, 3 and 20 hours and analyzed the DNA damage by the Comet assay. We found progressive dose-dependent degradation of nuclear DNA in exposed HK and dermal HF as a function of As dose and time of exposure (Figures 3.30–3.33). Hydrogen peroxide-exposed cells were used as a positive control. Cells exposed to only aqueous medium (MEM) were used as a negative control. The low concentration of 10 μg/L during short exposures (1 to 2 hours) caused only slight changes in DNA integrity, whereas longer exposures (20 hours) to the low concentration resulted in longer tails (Figure 3.30). Higher concentration of As(III) (500 μg/L) caused damage in a time span as short as 30-min exposure (Figure 3.31). Since virtually all of the cells exposed to 1000 μg/L developed tails at any duration of exposure used, we discontinued analysis of DNA damage produced in cells at this concentration.

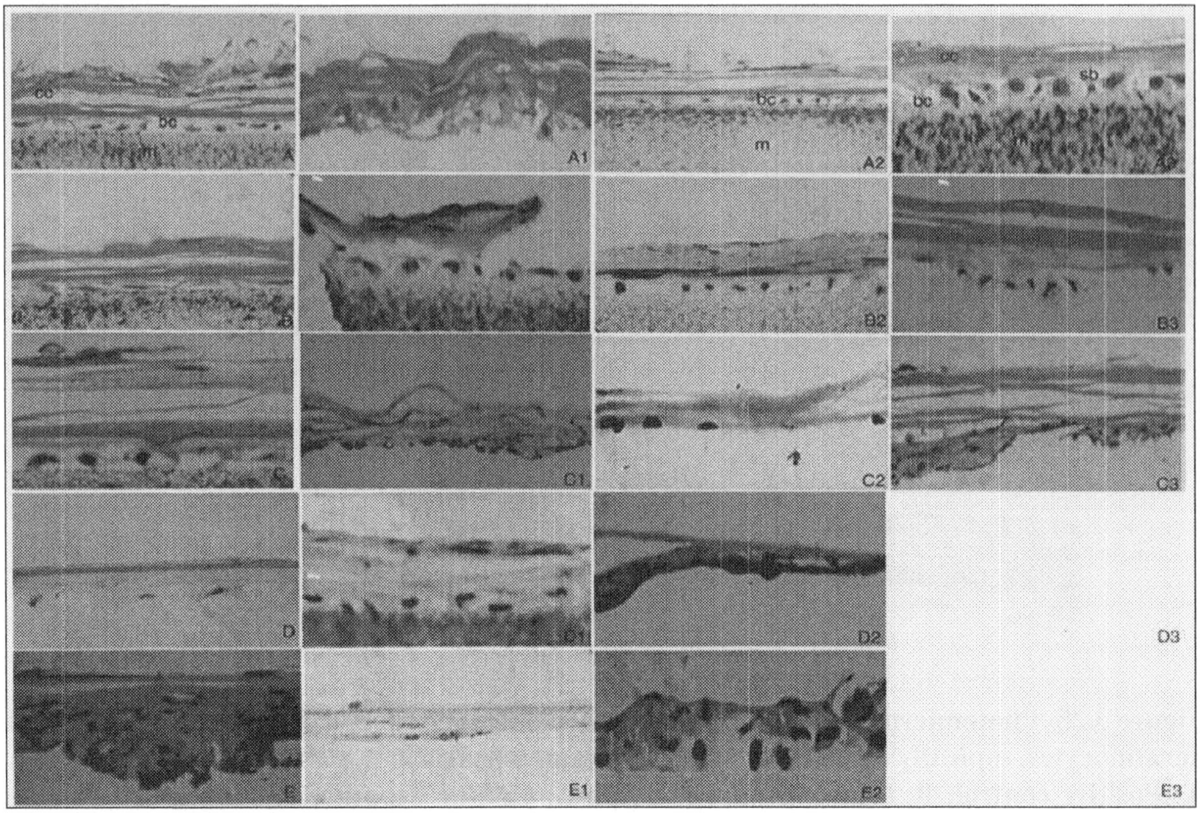

Figure 3.29 Photomicrographs showing dose-dependent changes in artificial skin, topically exposed to As(III) for 1, 6, 48 and 72 hr.
Horizontal rows represent time series: 1 through 72 hours (A through A3).
Vertical rows represent concentration series—A: control; B: 10 μg/L; C: 100 μg/L; D: 500 μg/L; E: 1000 μg/L. D3 and E3 not shown since the cultures were completed degraded. Additional note m = membrane, bc = basal cells, cc = cornified cells, sb = suprabasal cells. Also, the dark brown color of affected nuclei represent apoptotic or necrotic changes. The staining marked specific in situ DNA fragmentation at the single cell level.

Exposure of HF to 5 μg/L As(III) for 30 min resulted in an increase of 165% in the comet moment. At 10 μg/L, the increase was 195%, rising to 282% at 100 μg/L and 728% at 500 μg/L. These data indicate a strong dose-response relationship between the exposure of cultured human fibroblasts to As(III) and the progressive increase in extent of DNA single-strand breaks as measured by the Comet assay technique. Even a brief exposure to low concentration of As(III) (5 μg/L) was capable of producing a large increase (165% over control) in the observed DNA strand breakage (Figure 3.33). The Comet assay clearly demonstrated the higher sensitivity of HF to arsenic compared to HK (Figure 3.32), and confirmed our previous observations based on the TUNEL techniques. The lowest observed effect dose with this method was 5 μg/L for As(III).

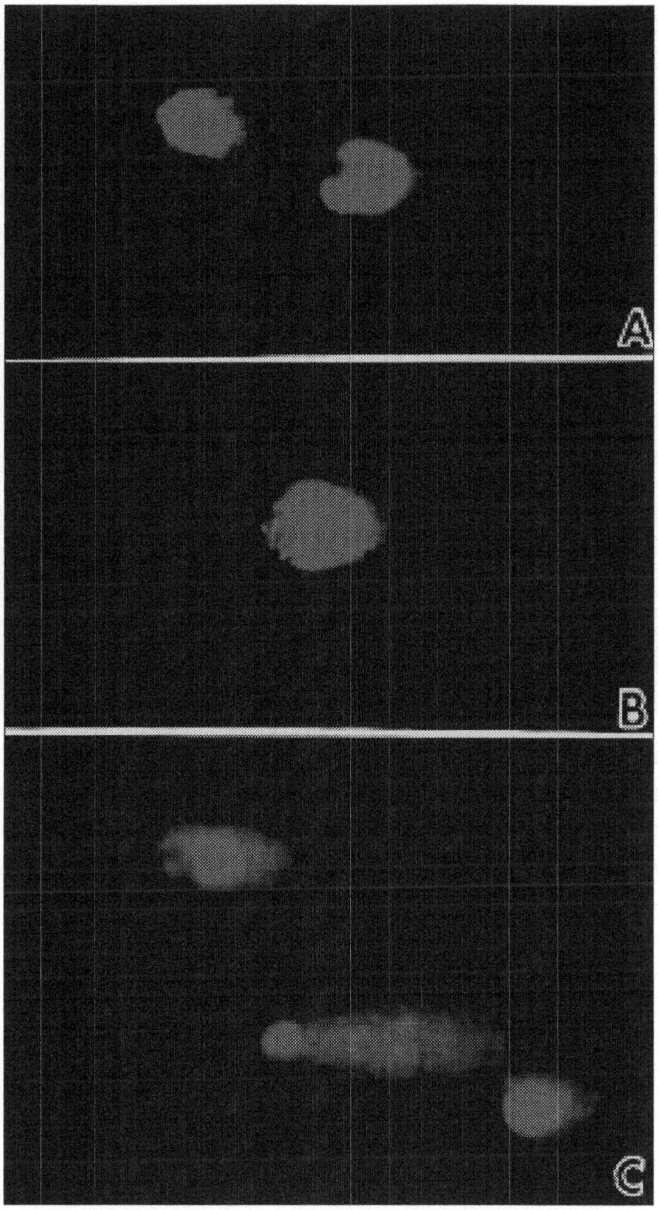

Figure 3.30 DNA damage in human keratinocytes following exposure to As(III). Cells, grown submerged for 24 hours, were exposed to 10 µg/L As(III), processed in the alkaline comet assay and viewed in an epi-fluorescence light microscope. A: 1 hr; B: 2 hours; and C: 20 hours after exposure to 10 µg/L of As(III). The photograph reflects time dependent DNA damage revealed by the presence of pronounced comet tails. Magnification ×400.

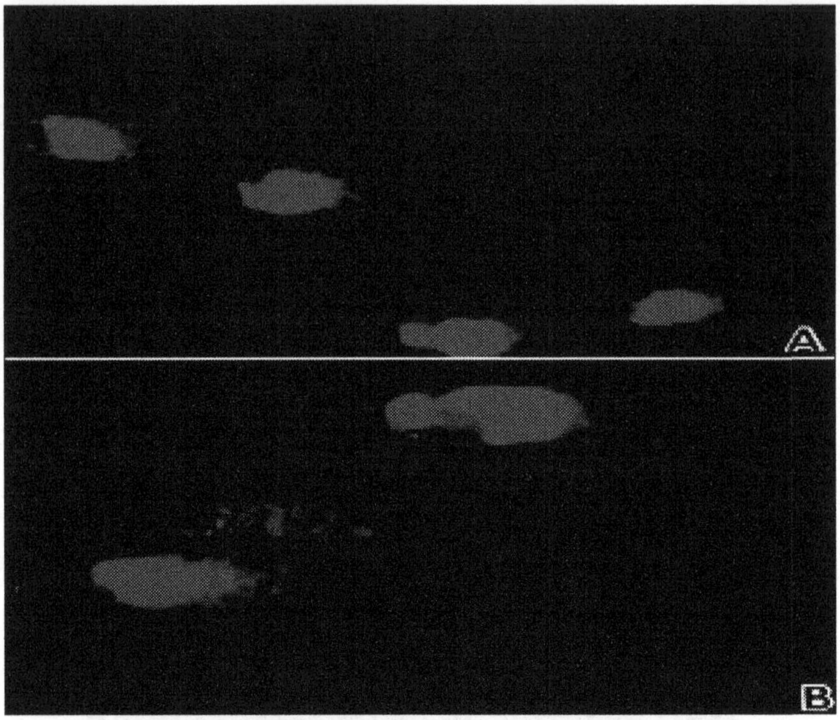

Figure 3.31 DNA damage in human keratinocytes following exposure to As(III). Cells, grown submerged for 24 hours, were exposed to 500 µg/L As(III), processed in the alkaline comet assay and viewed in an epi-fluorescence light microscope. A: 30 min; B: 2 hours after exposure to 500 µg/L of As(III). The photograph shows DNA damage depicted by the presence of pronounced comet tails. Magnification ×400.

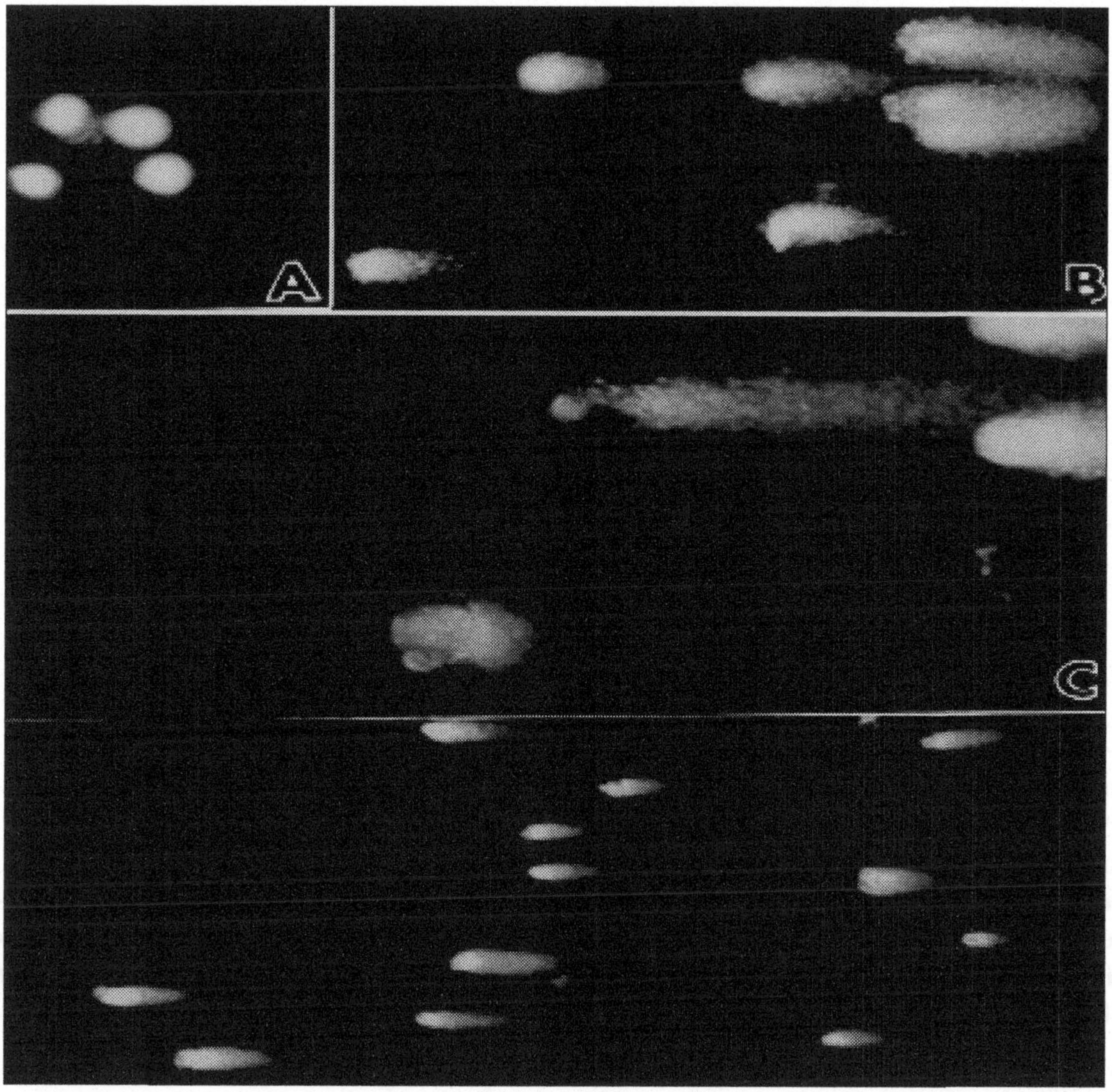

Figure 3.32 DNA damage in human fibroblasts following exposure to As(III). Cells, grown submerged for 24 hours, were exposed to 500 µg/L As(III), processed in the alkaline comet assay and viewed in an epi-fluorescence light microscope. A: control; B: 30 min; C and D: 1-hr after exposure to 500 µg/L of As(III). The photograph shows DNA damage presented as the pronounced comet tails in B–D. Magnification: ×400 (A–C) and ×100 (D).

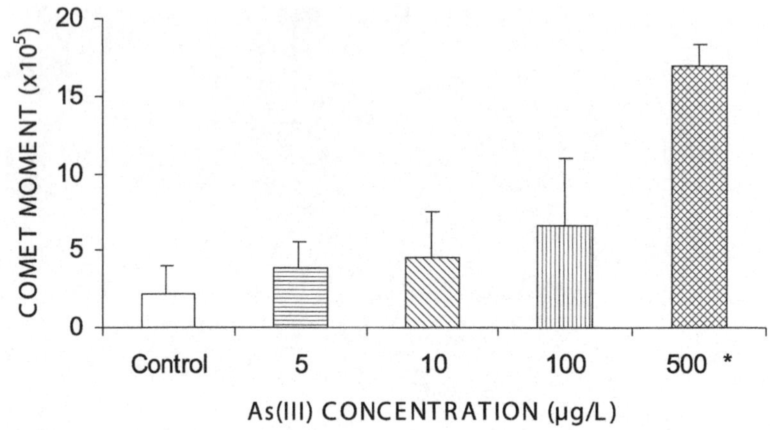

Treatment condition	Average comet movement ± S.D.	Min.	Max.
Control	$2.23 \times 10^5 \pm 1.74049$	41437	674558
5 µg/L	$3.83 \times 10^5 \pm 177841$	75398	808001
10 µg/L	$4.54 \times 10^5 \pm 288085$	152965	1459898
100 µg/L	$6.57 \times 10^5 \pm 437037$	213269	1812499
500 µg/L*	$16.95 \times 10^5 \pm 1355524$*	221056	5047236

* Denotes a fragment of dead tissue accumulated in the course of normal cell culturing not representative of tissue response.

Figure 3.33 Quantitative analysis of DNA damage in HF following exposure to As(III) for 30 min. The mean ±S.D. values of tail moments were calculated randomly for 20 cells per slide and three slides per sample. Level of significance was at $P<0.05$. Note: S.D. = standard deviation, Min. = minimum value of the tail moment measured, Max. = maximum value of the tail moment measured.

Evaluation of Apoptosis at the Chromosome Level (Telomerase Assay)

These experiments show that changing the exposure of HF to 100 and 1000 µg/L As(III) from 15 min to 30 min elevated the telomerase activity in both cases (Figure 3.34). The enhancement was greater at the higher 1000 µg/L As(III) concentration (~3 times control) and less pronounced at the lower 100 µg/L exposure (~1.6 times control), thereby reflecting the significant effect of As(III) dose on chromosome damage. This assay alone did not allow the determination of whether the enhanced telomerase activity reflected lasting or rapidly repaired chromosomal damage. However, taken together with our observations in the Comet assays, the degree of DNA fragmentation at these exposures (15-30 min at 10 and 100 µg/L) was consistent with global destruction of the nuclear chromatin structures at 1000 µg/L for 15 min as demonstrated by the increases in the comet tail moments.

It would be interesting to establish whether and to what extent the chromosomal DNA damage produced by such low level exposure (15-30 min at 10 and 100 µg/L) could be repaired.

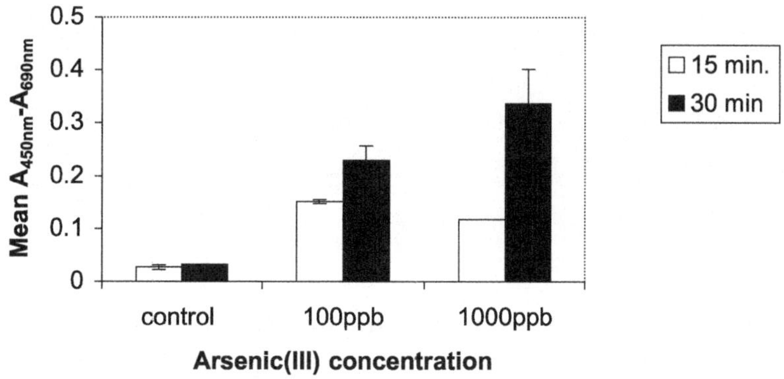

Treatment condition			
As(III) ppb	Exposure time	RNase	Mean $A_{450\,nm} - A_{690\,nm} \pm$ S.D.
Control	15 min	+	.027 ± .004
100 ppb	15 min	-	.151 ± .004
1000 ppb	15 min	-	.117 ± 0
Control	30 min	+	.032 ± .001
100 ppb	30 min	-	.229 ± .028
1000 ppb	30 min	-	.337 ± .064
Positive Control	-	-	3.815 ± .226

Figure 3.34 Telomerase activity measured by TRAP assay. Note: DNA free-RNAase was preincubated within cell extracts from the negative control group to completely destroy any expression for telomerase.

Chromosomal Damage

Human fibroblasts exposed to different doses of As(V) for 5-hr revealed a 5-fold increase in mitotic figures at 10 µg/L, 100 µg/L and 500 µg/L. At 1000 µg/L exposure, the detected increase was only 1.5-fold, apparently reflecting intense degradation of the exposed cells. The peculiarity of skin cells (such as cornification) as well as the type of their maturation, along with the decrease in mitotic activity in the older cultures impose certain restriction and limitations on the ability to isolate intact chromosomes for cytogenetic analysis. Attempts were made to develop various approaches to increase the yield of chromosomes but the lifted cultures yielded unsatisfactory chromosome preparations, while submerged cultures seemed more suitable of this objective (Figure 3.35).

EFFECTS OF ARSENIC ON SIGNAL TRANSDUCTION (WESTERN BLOT METHOD)

Western Blot technique was employed in the analysis of the expression of heat shock proteins 70 (HSP70) in HK exposed to 10, 100 and 500 µg/L As(III) for 6 days. A dose-dependent increase in HSP70 expression was observed (Figure 3.36A) Exposure to 10 µg/L resulted in

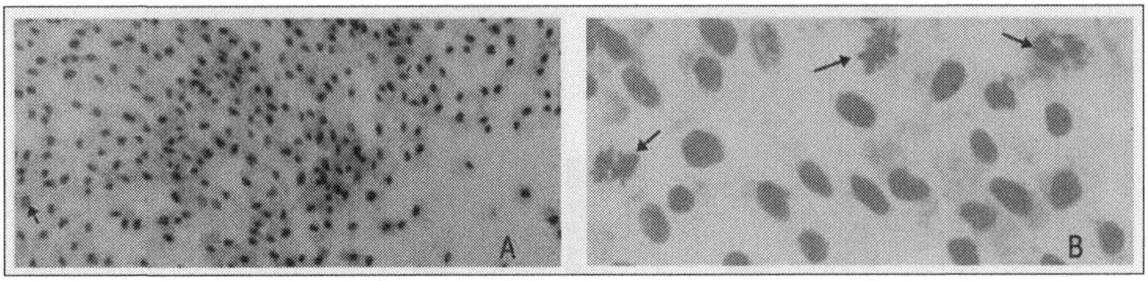

Figure 3.35 Chromosomal damage in human fibroblasts following exposure to As(III).
A: control; B: three hours after the cultures were exposed to 10 µg/L of As(III). Arrows point to damaged chromosomes. Magnification ×130 (A), ×460 (B).

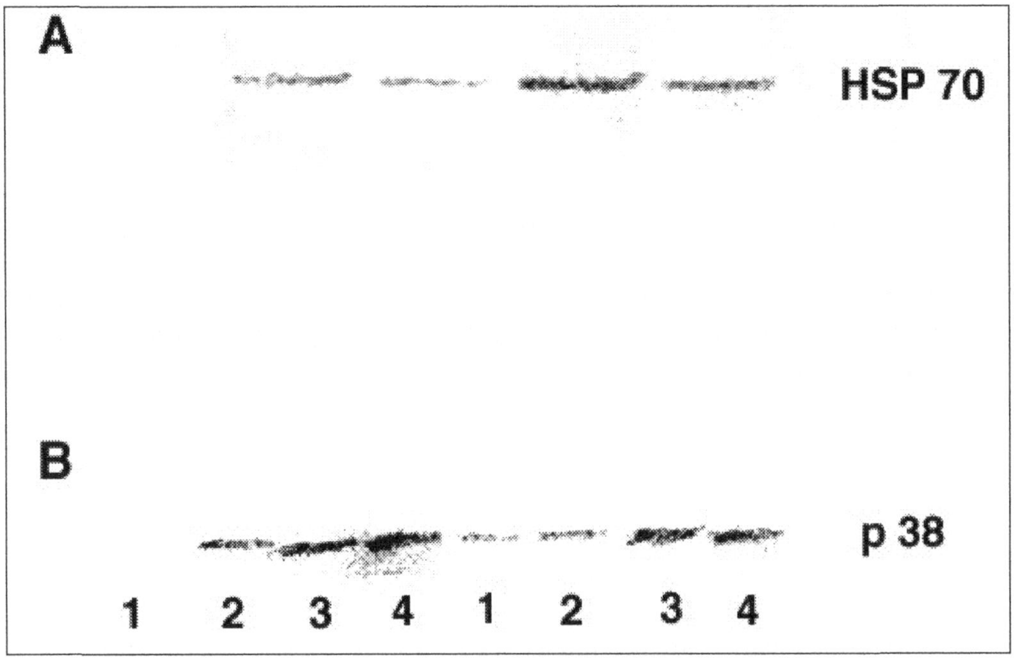

Figure 3.36 Western blots of HSP-70 and p38 reactive proteins isolated from submerged cultures of HK, exposed to As(III).
A: Western blots of HSP-70 reactive proteins. Numbers indicate 1: control; 2: 10 µg/L; 3: 100 µg/L; and 4: 500 µg/L of As(III).
B: Western blots of p38 reactive proteins. Numbers indicate 1: control; 2: 5 µg/L; 3: 10 µg/L; and 4: 500 µg/L of As(III).

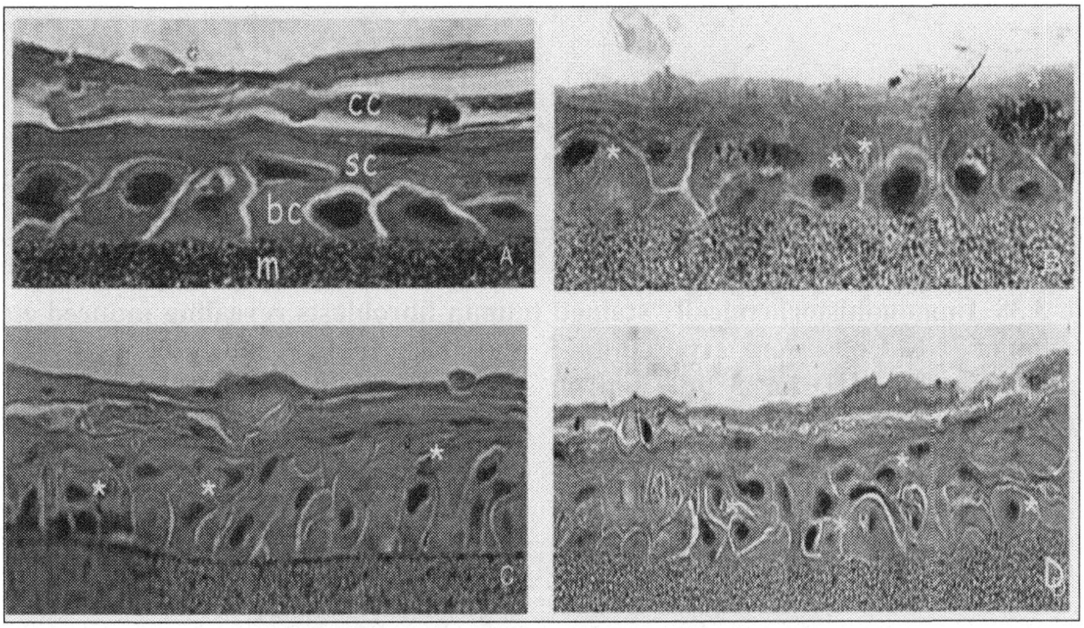

Figure 3.37 Immunohistochemically stained artificial skin revealing p53-immunoreactive keratinocytes. A: control; B: 6 hours at 100 µg/L; C: 48 hours at 100 µg/L; and D: 6 hours at 500 µg/L. m = membrane, bc = basal cells, sc = spinous cells, cc = cornified cells, * = p53 positive cells. Asterisk indicates areas of deposits indicating the presence of the specific antigen (p53 protein).

HSP70 band value of 0.80 while at 100 µg/L this value was found to be 4.06. The increase should also be considered in the context of progressive loss of protein in HK exposed to these As(III) doses. Taking into account the protein loss of some 45% at 100 µg/L (see Figure 3.20), the increase in the expression of HSP70 in these experiments was estimated to be 5.89, i.e., over 7-fold.

A similar dose-dependent increase in the expression of the signal transduction protein p38 was observed in HK exposed to 10 µg/L As(III) for 6 days (Figure 3.36B).

GENOTOXICITY EXPERIMENTS

Paraffin sections of cultures stained with H&E revealed noticeable morphological changes in cells exposed to As(III), similar to what had been reported in previous sections. Exposure to 100 µg/L of As(III) for six hours was found to cause a marked increase in p53 expression (Figure 3.37B) in basal and spinous cell layers. The p53 staining was evident in keratinocytes and early spinous cells in cultures exposed to 100 µg/L for 48 hour (Figure 3.37C) and to 500 µg/L for 6 hours (Figure 3.37D).

Exposure of both HK and HF to arsenic at the low doses (5 µg/L for 1-3 hrs) produced activation of early stress-related and mitogen activated kinases. We tested the response of cells in two culture formats, i.e. grown as submerged cultures or cultured as artificial skin on membranes. Submerged cultures (Figures 3.38 and 3.39) showed more intense staining at 5 µg/L than artificial skin (Figure 3.40).

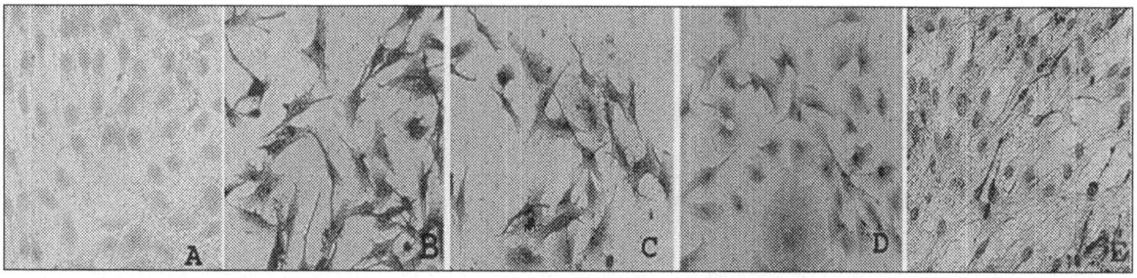

Figure 3.38 Immunohistochemically stained human fibroblasts revealing induced elevated expression of JNK. Cells were exposed to 5 µg/L (B), 10 µg/L (C), 100 µg/L (D), and 1000 µg/L (E) of As(III) for 3 hr. Control: A. Dark gray deposits indicate the nuclear presence of the activated JNK in B–E.

Figure 3.39 Immunohistochemical staining of human keratinocytes exposed to 100 µg/L of As(III) for 3 hr. Dark gray deposits indicate the presence of activated p38 in A and JNK in B. Note higher expression of p38 in (A) than of JNK in (B).

It is noteworthy that elevated expression of p38 and JNK can be seen in HK exposed to the lowest dose of As(III) of 100 µg/L for 3 hr (Figure 3.39 A and B)

Comparison of the degree of expression of p38 and JNK in HF and HK suggest that the signal transduction pathway response in HF is somewhat stronger and is detected at lower As(III) doses that in HK. These are initial observations that would have to be confirmed in more extensive and detailed experiments. However, the greater sensitivity of HF compared to HK was noted previously in these studies as well as in DNA fragmentation studies by the Comet assay.

In both cases (submerged cultures and artificial skin), the sensitivity of cells to arsenite was similar, demonstrating marked activation of p38, ERK and JNK at the lowest arsenate levels as early as within 1-3 hrs of exposure. The extent of activation of these stress markers was increased in a dose-dependent manner so that progressively higher number of cells showed positive staining (Figures 3.38–3.40). The dose dependent activation of these signal transduction components correlates with the progressively enhanced cyto- and genotoxicity of higher doses of As(III).

At the same dose and time of exposure, as shown in Figure 3.40, arsenite induced p38 activity (A–C) to a higher level than that of ERK (D–F) and JNK (G–I). This phenomenon was also reported by Ludwig et al. (1998). Activation of JNK in submerged cultures of

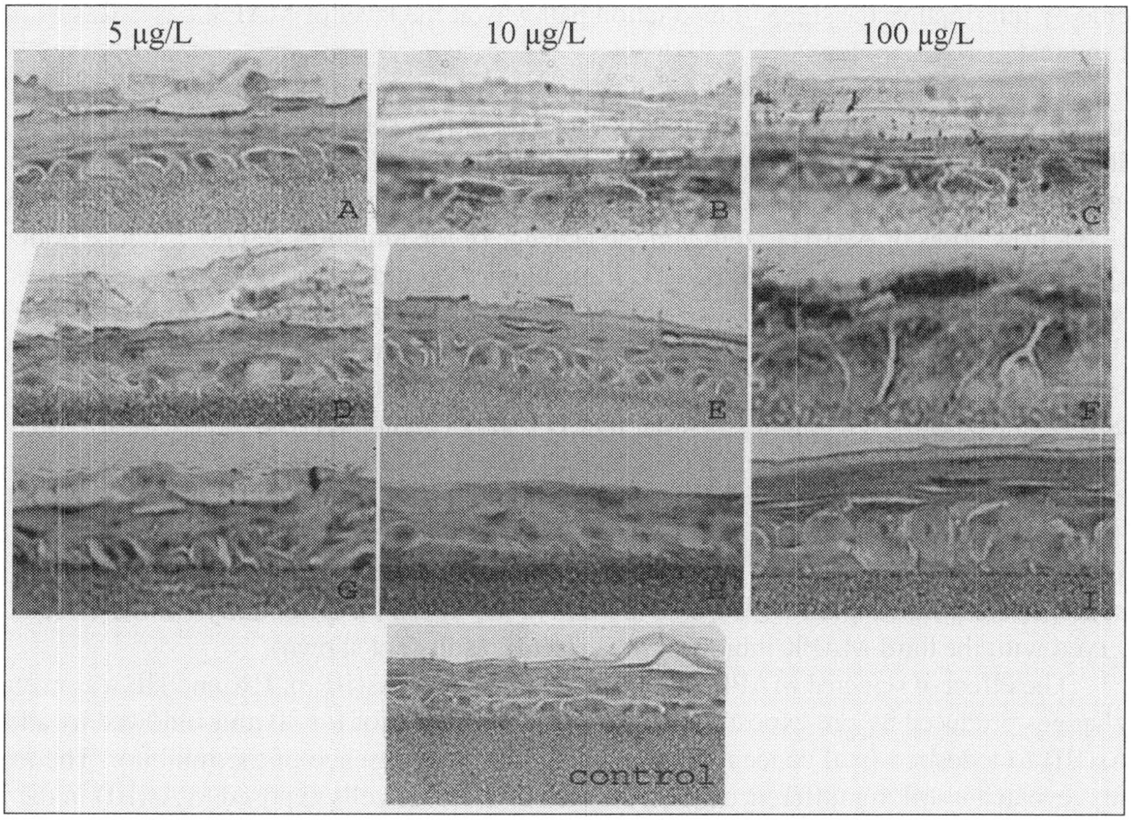

Figure 3.40 Immunohistochemically stained artificial skin topically exposed to 5 µg/L (A, D, C), 10 µg/L (B, E, H), 100 µg/L (C, F, I) and control. First row = p38, second row = ERK, third row = JNK. Dark gray deposits indicate the presence of activated p38 in A, B, C; ERK in D, E, F; and JNK in G, H, I. Note higher intensity of nuclear and cytoplasmic staining of p38 in C (100 µg/L) compared to that in A (5 µg/L). Note the relatively light ERK activation (second row) in upper differentiated layers at 5 and 10 µg/L and very pronounced activation of ERK in the upper layer at 100 µg/L. The control (unexposed cells) do not show any activation of p38, ERK or JNK.

human keratinocytes and fibroblasts (Figures 3.38–3.39) was much stronger than in artificial skin (Figure 3.40).

INHIBITOR EXPERIMENTS AND PREVENTION OF ARSENIC TOXICITY

A number of compounds are known to inhibit the toxicity of arsenic including *SB 202190* [4-(4-Fluorophenyl)-2-(4-hydroxyphenyl)-5-(4-pyridyl)1H-imidazole; $C_{20}H_{14}FN_3O$]; *SB 203580* [4-(4-Fluorophenyl)-2-(4-hydroxyphenyl)-5-(4-pyridyl)1H-imidazole; $C_{21}H_{16}N_3OSF$]; *SB 202474* [4-Ethyl-2(p-methoxyphenyl)-5-(4'-pyridyl)-IH-imidazole; $C_{17}H_{17}N_3O$]; *PD 98059* [2'-amino-3'-methoxyflavone($C_{16}H_{13}NO_3$); *Nimesulide* [NS], and *Niflumic acid* [NA]. The last phase of this project included preliminary experiments on mechanisms of inhibition of arsenic toxicity by these compounds. The ability of specific inhibitors to counteract the cytotoxic and genotoxic effects of arsenic by interrupting the signal transduction pathway at the level of MAP kinases was of particular interest.

Phase-contrast observation of cultures grown for 4 days in the presence of different concentrations of As(III) and MAPK inhibitors (SB 203580 and PD 98059) reveal that the MAPK inhibitors alone (at the concentration used in this study) did not produce any morphological changes compared to the control cultures as shown in the first row of pictures (Figure 3.41A, 1A and 2A). Low doses of As(III) (5 μg/L) in the absence of the inhibitors did not produce noticeable morphological changes although at 10μg/L slight changes could be detected in some cultures. The presence of the inhibitors did not alter this picture at these As(III) concentrations. At 100 μg/L nuclear abnormalities as well as overall enlargement of cells could be seen (see second row of pictures in Figure 3.41B. This damage was completely prevented by MAPK inhibitor SB 203580 (Figure 3.41 1B) and even to a large extent by PD 98059 (Figure 3.41 2B).

An even more protective effect of the MAPK inhibitors could be seen at the higher concentration of 500μg/L As(III) (see Figure 3.41, C–control, 3.41, 1C - SB 203580, 2C - PD 98059). At 1000 μg/L of As(III), one can see virtually complete destruction of cells in the absence of the MAPK inhibitor (Figure 3.41D), and a strikingly preserved morphology of the same cells in the presence of SB 203580 and PD 98059 (Figure 3.41 1D and 2D). Essentially similar effects were observed with the third MAPK inhibitor SB 202190 (results not shown).

The effect of selected MAPK inhibitors on toxicity of As(III) in HK and HF was evaluated by changes produced by pre-exposure of cells to a given inhibitor for 30 min followed by addition of As(III) to a desired final concentration in the continued presence of the inhibitor. The experiments revealed a striking difference in appearance of cultured cells exposed to As(III) in the presence and absence of the MAPK inhibitors SB 203580, PD 98059 and SB 202190. Specifically, high level of expression of p38 in HK exposed for 3 hours to 100 μg/L of As(III) noticed in the cytoplasm and in the nucleus (Figure 3.42A). Preincubation with MAPK inhibitor SB 203580 blocks the expression of p38 (Figure 3.42B). Comparison of dose-related changes in cell morphology and the level of expression of p38 of HF show the protective effect of SB 203580 at the lower level of exposure to As(III) (5 and 10 μg/L for 2 hr). This is manifested in better preservation of cellular morphology and a markedly lower activation of p38 (Figure 3.43 1B and 1C). At higher As(III) concentration (100 μg/L, 2 hr) HF without the inhibitor display significant deterioration of cellular morphology whereas SB 203580 virtually prevented any visible damage at the same HF exposure dose that was accompanied by a marked reduction in the expression of p38 (Figure 3.43D and 1D).

Likewise, at higher As(III) exposure dose of 500 μg/L, the destruction of cytoplasmic and nuclear integrity is largely prevented by SB 203580 (Figure 3.43 1E). The same protection is seen in cells exposed to the highest tested concentration of As(III) of 1000 μg/L where cellular morphology in the absence of the inhibitor is totally deteriorated. In stark contrast, cells exposed to 1000 μg/L As (III) in the presence of SB 203580 not only retain their morphological features but apparently retained a significant degree of functional integrity as well, since one can observe significant expression of p38 largely confined to the nucleus in these cell (data not shown).

It is known that p38 initially located in the cytoplasm, upon activation, moves to the nucleus where it triggers transcriptional activation that constitutes genomic response to a stress signal. It is apparent, however, that at high As(III) level (1000 μg/L), a degree of alteration in normal cellular morphology does take place; the extent of microscopically detectable damage is much less than the usually observed total cellular lysis in cells unprotected by the SB 203580.

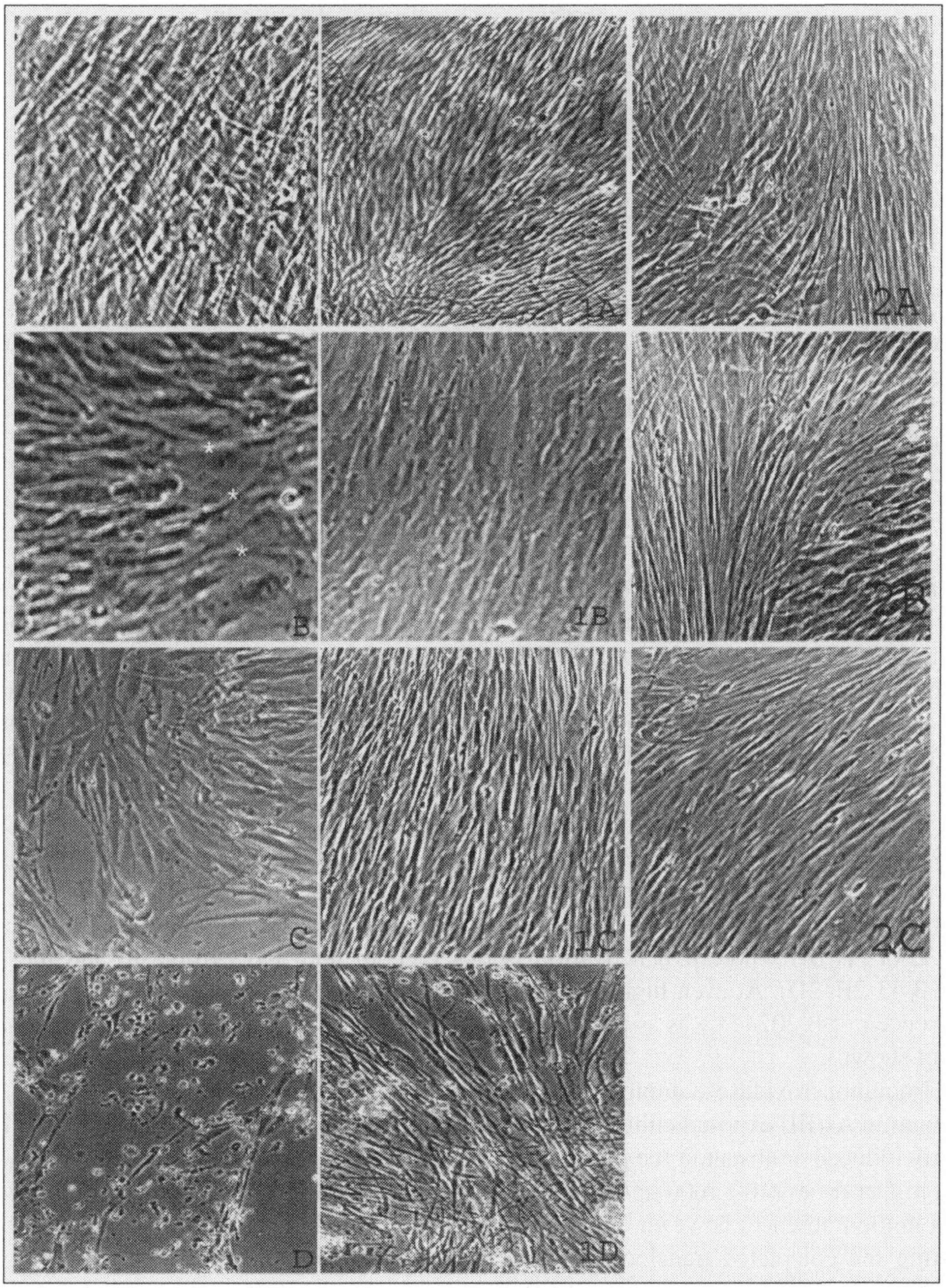

Figure 3.41 Phase contrast micrographs of human fibroblasts exposed to different concentrations of As(III) for 4 days alone (first column) and in the presence of MAPK inhibitors SB 203580 (second column) and PD 98059 (third column). First row = control (no As(III) added), second row = 100 μg/L, third row = 500 μg/L, and fourth row = 1000 μg/L As(III) added.

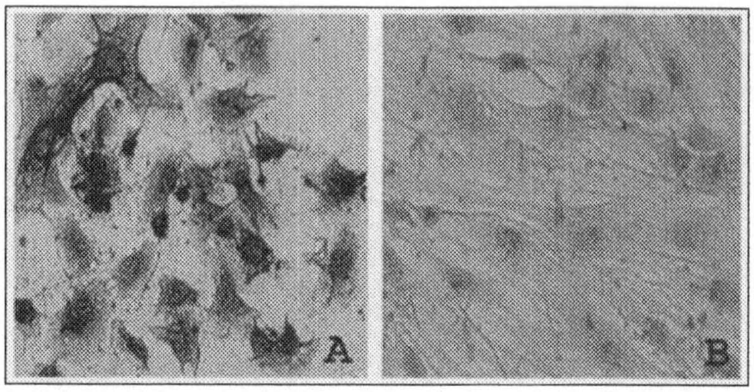

Figure 3.42 Effect of MAPK inhibitor SB 203580 on immunoperoxidase staining of p38 in human keratinocytes exposed to As(III). A: Cells exposed to 100 μg/L of As(III) for 3 hr. Dark gray deposits indicates the cytoplasmic and nuclear presence of the activated p38. B: Cells exposed to 100 μg/L of As(III) for 3 hr in the presence of MAPK inhibitor SB 203580. Note good morphology and absence of activated p38.

Essentially similar findings were made with the other MAPK inhibitors PD 98059 and SB 202190. It appears, however, that PD 98059 provides somewhat less protection against activation of p38 than SB 203580 at the same As(III) concentrations (Figure 3.43 2A–2B). It should be noted that PD 98059 also confers remarkable protection against morphological deterioration produced by As(III) toxicity. At the same time, however, it is less effective than SB 203580 in reducing the activation of p38 while being effective in inhibiting activation of other MAP kinases. It is likely, therefore, that the protective action of PD 98059 against morphological destruction by As(III) is achieved largely through inhibition of MAPKs other than p38.

In order to confirm the specificity of action of MAPK inhibitors SB 203580, PD 98059 and SB 202190, a compound SB 202474 that has a similar structure but lacks MAPK inhibiting activity was used as a negative control. We find that incubation of cells with SB 202474 did not prevent p38 expression in As(III) exposed cells; this fact is strongly evident in HF exposed to 500 μg/L of As(III) where p38 activation is observed in both the cytoplasm and the nucleus (Figure 3.43 2B–2D). At even higher concentration of 1000 μg/L As(III), the cell destruction in the presence of SB 202474 was as pronounced as in control cells incubated without any inhibitors (data not shown).

Immunoperoxidase staining for p38 shows elevated expression of this signal transduction component in As(III) exposed cultures without MAPK inhibitors (Figure 3.43 B–D) which is significantly reduced or absent in the presence of the MAPK inhibitors (Figure 3.43, frames 1B, 1C, 2B). At higher doses (100-1000 μg/L), the morphology of cells in the presence of inhibitors is better than in their absence; however, the expression of p38 is still apparent in the nuclei of the cells suggesting the completed translocation of activated form of p38 to the nucleus (Figure 3.43, frames 1D, 1E, 2C, 2E). The presence of SB 202474 MAPK negative control added to As exposed cells at 100 μg/L revealed significant expression of p38 and marked destruction of cellular morphology (Figure 3.43 negative control).

MAPK inhibitor SB 203580 provides significant protection against morphological changes and activation of JNK in extended (24 hr) exposure to higher concentrations of As(III) (100 and 500 μg/L) (Figure 3.44, frames A, C, E—SB 203580; B, D, F—PD 98059; G—Control in the absence of inhibitors).

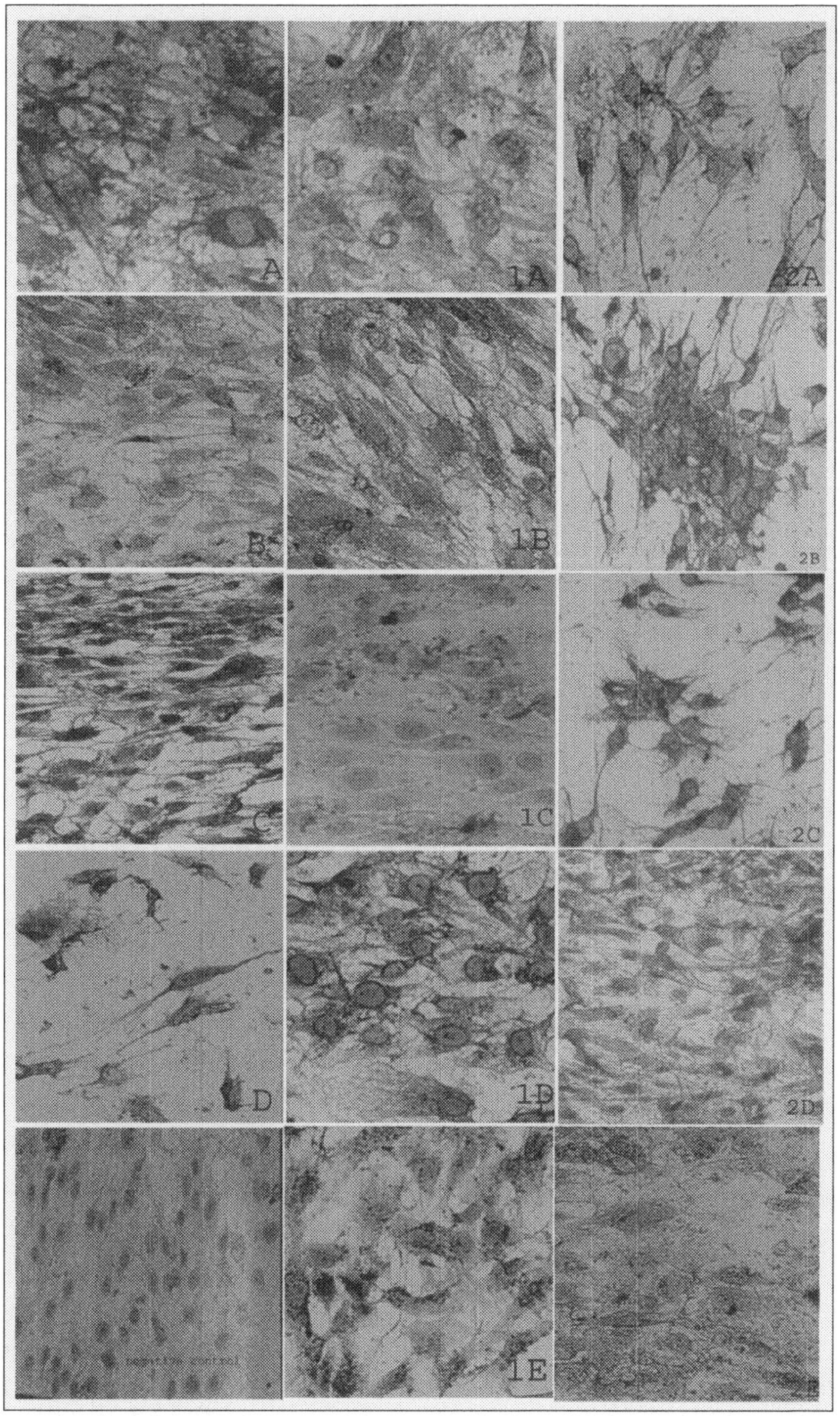

Figure 3.43 Photomicrographs of immunochemical assays for p38 in HF exposed to different concentrations of As(III) for 2 hours alone (first column) and in the presence of MAPK inhibitors SB 203580 (second column) and PD 98059 (third column). First row = control (no As(III) added), second row = 5 μg/L, third row = 10 μg/L, fourth row = 100 μg/L, and fifth row = 500 μg/L As(III).

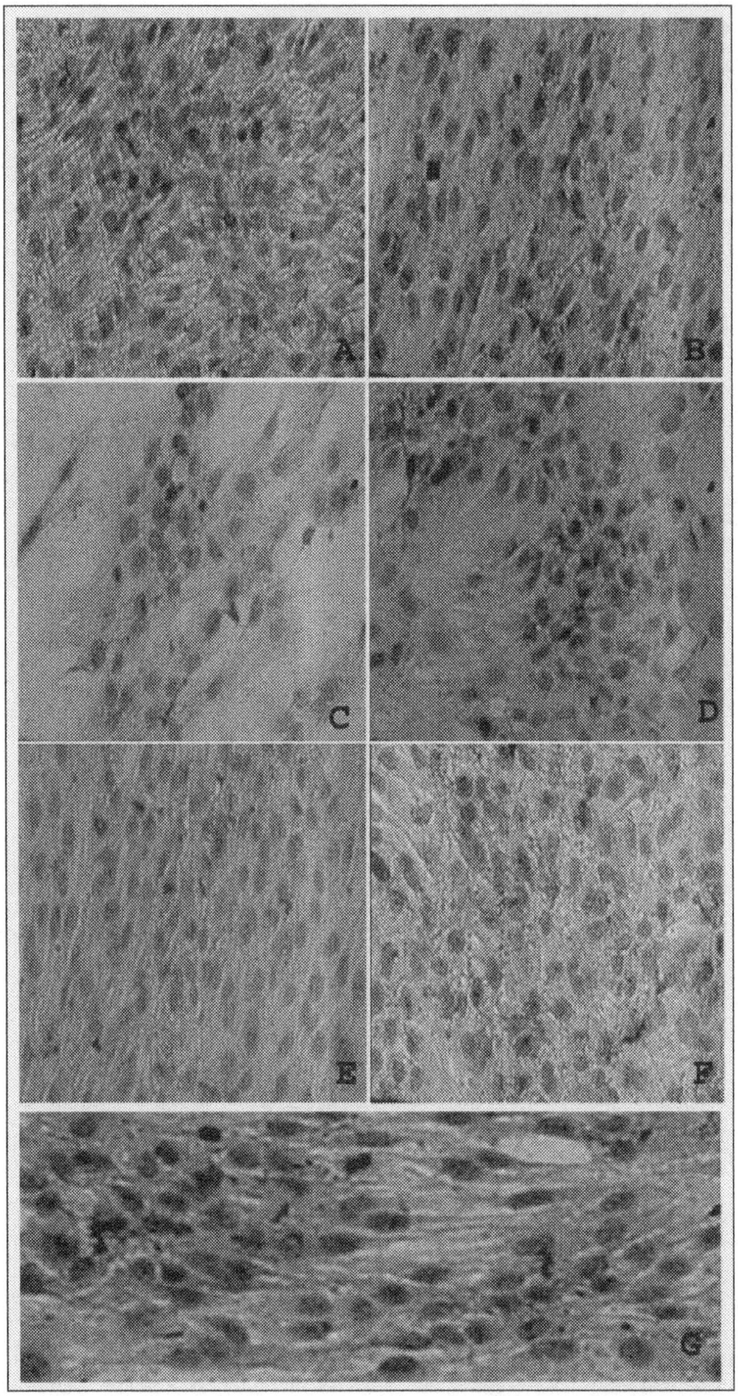

Figure 3.44 Photomicrographs of HF exposed to different concentrations of As(III) for 24 hr and immunochemically stained for JNK in the presence of MAPK inhibitors SB 203580 (A, C, E) and PD 98059 (second column). First row = 10 µg/L As(III), second row = 100 µg/L As(III), third row = 500 µg/L, and fourth row = control no As(III) added.

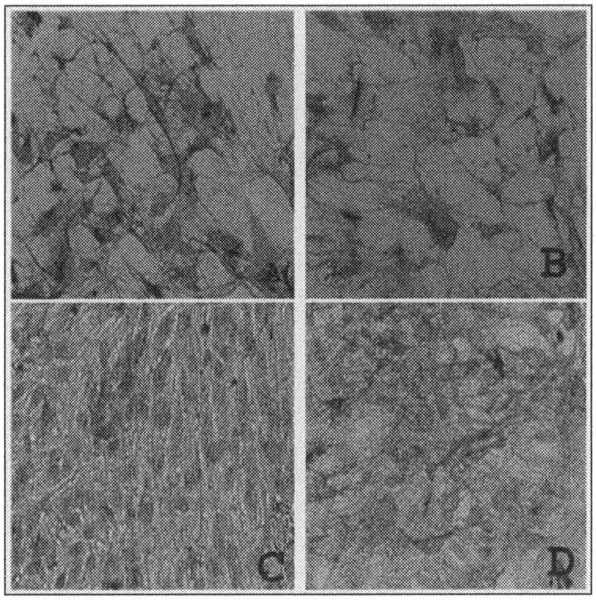

Figure 3.45 Photomicrographs of immunochemical staining for p38 expression in HF exposed to As(III) in the presence of COX-2 inhibitors NS-398 (A: 500 μg/L; B: 1000 μg/L) and Niflumic acid (C: 500 μg/L; D: 1000 μg/L).

In the presence of the COX-2 inhibitor NS-398, the p38 activity was somewhat reduced and preservation of cellular morphology was seen mostly at the higher As(III) exposures (500 and 1000 μg/L) (Figure 3.45). The effect of Niflumic acid at these As(III) concentrations is less pronounced than that of NS-398 and, as expected, the cells displayed abundant p38 expression and a greater loss of morphological integrity (Figure 3.45 C, D). The effect of Nimesulide was almost identical to that of Niflumic acid (results not shown).

It appears that some COX-2 inhibitors provide some degree of protection against morphological destruction by As(III); this is achieved without significantly influencing expression of p38 in the signal transduction pathway. The COX-2 inhibitor Niflumic acid provided virtually no protection against morphological damage produced by As(III) at any concentration tested (Figure 3.45 C and D). Although JNK expression appears to be somewhat depressed by Niflumic acid, conclusive evidence to support this impression could not be obtained.

In summary, the preliminary studies show that the COX-2 inhibitors tested, namely, NS-398, Niflumic acid and Nimesulide provided less protection against the destructive effects of As(III) exposure than the protection provided by any of the three MAPK inhibitors, namely, SB 203580, PD 98059 and SB 202190 tested. This is manifested both at the level of expression of p38 and JNK activation as well as in the level of morphological integrity in As(III)-exposed HF. The preservation of cellular structures when specific MAPK inhibitors were presented to As-exposed cells would suggest that in order to counteract the destructive cytotoxic and genotoxic effects of As(III) exposure, the application of specific inhibitors able to interrupt the signal transduction pathway at the level of MAP kinases is more efficient than disruption of reactive prostaglandin synthesis induced by toxic metal exposure. This study points to a novel protective and preventive approach for mitigating chronic and acute As toxicity using specific inhibition of the signal transduction pathway. The reduction in arsenite toxicity may be attenuated by using selective agents (such as the COX-2 inhibitors) of the reactive branch of the cellular stress response mechanism represented by activation of the inducible prostaglandin synthesis. The details of specific components in this pathway that can serve as the most appropriate targets for such inhibitor action will have to be studied further. To the best of our knowledge, these findings present the first experimental demonstration that specific inhibition of signal transduction pathway may offer practical benefits in preventing cytotoxic and genotoxic effects of acute and chronic As exposure in susceptible populations.

CHAPTER 4
DISCUSSION

GENERAL

The affinity of arsenic for keratinizing structures makes the skin a critical organ in arsenic toxicity. Systemic toxicity becomes manifested in a variety of symptoms including keratosis, hyperpigmentation, skin eruptions, dermal lesions (such as arsenical epithelioma, verrucous hyperkeratosis, warts and melanosis), Bowen's disease (precancerous dermatitis), various forms of skin cancer, and neoplasia (Guy et al., 1999). Chronic dermal accumulation of arsenic conceivably increases the susceptibility of the skin to ultraviolet radiation. With latency period of up to 30 years (Leyh and Rothlaender, 1985), the skin lesions can be manifested long after the arsenic exposure has ceased and systemic levels have returned to normal (Luchtrath, 1983).

This project has examined the effects of As exposure on human skin at four levels of functional cellular organization: morphological and ultrastructural assessment of the damage (gross and microscopic changes, loss of selective permeability), generalized biosynthesis processes (DNA, RNA, protein synthesis), activation of selected stress-induced signal transduction components (HSP70 and p38) and genotoxic effects reflected in DNA fragmentation (TUNEL, COMET, Telomerase assays and expression of p53 protein). The studies are based mainly on lab-grown human skins which allowed a higher degree of standardization than could be expected from random *ex vivo* human skin biopsy material. These artificial skin cultures develop well ordered stratification and extensive cornification similar to normal human skin (Bernstam, Vaughan, and Bernstein, 1986; Bernstam et al., 2002). There are, however, some differences between the natural skin and any culture system. For instance, epidermal desquamation is unlikely to occur in culture as is often the case *in vivo*, and there is no active vascular system in a skin culture. While donor age-specific molecular markers have been identified in dermal fibroblasts and keratinocytes, gene expression patterns in cultured dermal cells reflect aging of the cells in culture rather than the donor age (Cristofalo et al., 1998). This view on the age-of-donor effect is also supported by numerous observations suggesting that responses of cultured dermal cells derived from newborn and adult rats showed similar permeability characteristics and resistance to various toxins (Scavarelli-Karantsavelas et al., 1990; Pu et al., 1995; Bernstam et al., 2002). In spite of these limitations, the artificial skins offer several unique features for experimentation. For instance, a toxic agent can either be applied topically on the surface of the culture or added to the culture growth medium to reach the basal cells from substratum. The two routes of entry of toxic compounds into the skin system (topically via showering, bathing and hand-washing versus reaching the basal cells from beneath, resembling the *in vivo* route of entry after ingestion of contaminated water and/or food) can thus be studied.

PERMEATION AND RETENTION OF ARSENIC BY HUMAN SKIN

The association of arsenic poisoning with development of hyperpigmentation, keratoses and skin cancer has been repeatedly documented in chronic poisoning from drinking well water polluted with arsenic (Jaafar et al., 1993; Kovacs et al., 1996; Ohnishi et al., 1997). Some hyperkeratotic lesions and squamous cell carcinomas *in situ* are reported which may progress to invasive carcinoma; other invasive squamous cell carcinomas will develop *de novo* (Shannon and

Strayer, 1989). These cutaneous squamous cancers may metastasize. In spite of the fact that the skin is the largest organ in the human body, it is remarkable that dermal absorption is generally ignored as a pathway of exposure to arsenic. The strong affinity of arsenic for keratinizing structures of the skin is well known (Guy et al., 1999). The strong propensity for retaining As(III) and As(V) is linked to the fact that the skin is a critical organ for arsenic toxicity. Dermal penetration may explain why Blackfoot disease is rampant in Taiwan, where rice farmers' feet are constantly in contact with groundwater, but not in other areas, such as India and Bangladesh, where water sources are similarly contaminated.

This project has investigated the differences in the uptake and retention of As(III) and As(V) which are probably related to the structural chemistry of these species. At physiological pH, arsenate is a charged oxyanion thought to enter the cells through existing pathways by mimicking the phosphate ion. By contrast, As(III) exists as an uncharged species, $As(OH)_3$ which can diffuse passively into cells. Within the cell, As(III) quickly binds to proteins that contain sulfhydryl groups whereas As(V) inhibits enzymatic reactions in which phosphate esters are formed and must first be reduced to As(III) before the binding with proteins. Easier penetration and binding apparently is related to much higher toxicity of As(III) compared to As(V), as documented in several experiments of this study.

This project demonstrates that cultured skins have a remarkable affinity for both As(V) and As(III). Little of the 10-1000 μg/L arsenic loaded onto the artificial skin could be detected in the effluent through the skin membrane after up to five hours (Table 3.1). To confirm that the arsenic was retained in the keratinocytes, two possible sources of uncertainty in the experimental set-up had to be addressed: (1) Did the artificial skin serve as a barrier to arsenic ions? If so, the differential separation and penetration of water molecules into the lifted culture would increase the concentration of arsenic in the residual solution. Analysis of the residual solutions did not show a change in arsenic concentrations. (2) Did the nylon support membrane absorb some of the arsenic? Effluents from solutions containing 50 and 500 μg/L through a Nylon bed of 1, 5, and 10 membranes (with no artificial skin) were analyzed. The retention of arsenic was found to be less than 1% (within experimental uncertainty) of the initial arsenic concentration. Since As(III) is expected to be present as neutral species at the pH conditions of the experiments, it would seem unlikely that this species is being retained by adsorption unto membrane surfaces. Indeed, the higher retention rate for As(III) by the skin layers used in the experiments (compared to the rate for As(V)) may be regarded as a de facto evidence that this species diffuses into cells passively; pentavalent arsenic should be adsorptively retained in the outermost skin layers compared to neutral As(III) species.

Changes in concentrations of As (III) and As(V) in hourly aliquots with different initial starting concentrations of As(III) and As(V) were measured (Tables 3.1 and 3.2). The breakthrough times for the skin cultures during these experiments varied from one hour for As (V) to two hours for As (III). After the initial conditioning phase, the retention attained a fairly steady state marked by constant concentration of arsenic in the collection solution. The observed concentrations were much less than the expected concentrations (15.4 μg/L for As(III) and 30 μg/L for As(V)) even after 6-hr exposure, implying substantial retention of arsenic even under the steady state conditions. Saturation of skin cultures with arsenic was not achieved after six hours (expected concentrations were not attained), suggesting that the skin culture system has a high retention capacity for arsenic. From the data for the permeation experiments, it is estimated that about 30% of As (V) passing through the samples was retained compared to over 90% of the As(III) which was retained. Previous studies reported that, on average, 60-90% of the absorbed arsenic is retained in the skin (NAS, 1999), which is consistent with the results of the present study.

For artificial skin, the absorption rate for arsenic is estimated to be about 6-8%/hr of the applied dose for As(III) and 1-8%/hr of the applied dose for As(V) (see Table 3.1). Our results are higher than the 2-6% (of applied dose) in 24 hours reported by Wester et al. (1993) for radiolabelled arsenate topically applied to human skin or skin of rhesus monkeys in vitro. Our results, however, are comparable to the absorption rate of 30% in 24 hours found for the dorsal skin of mice (Rahman, Hull, and Hughes, 1994). A controlled study on rate of absorption of inorganic arsenic through intact human skin has not been done on a systematic basis. From the results of our experiments, the permeability constant (Kp) for arsenic can be calculated according to the following equation:

$$K_p(\text{cm/hr}) = \frac{Q}{(A \times C \times t)}$$

where Q us the total absorbed dose in microgram, A is surface area exposed (0.32 cm^2 for the setup used in this study), t is time in hours and C is concentration gradient across the skin membrane. The calculated Kp values averaged 4.3×10^{-5} cm/hr for As(V) with exclusion of the value for the 1000 μg/L dose and 10.1×10^{-5} cm/hr for As(III) (Table 3.2). These values are higher than the 2.7×10^{-6} reported for percutaneous absorption of As(V) by rat skin (Dutkiewicz et al., 1977), but lower than the default value of 1.0×10^{-3} cm/hr suggested by USEPA (1992) for percutaneous absorption of inorganics into human skin.

The results on percutaneous absorption can be used to estimate, in a rather rough way, the dermal absorption of arsenic during showering activity. Assuming that the thickness of the film of water in contact with the skin surface during showering is 100 μm, and from the determined percentage of applied dose absorbed, the absorption of As can be estimated per sq. cm of the body surface (Table 3.2). Using average surface areas for adult (20,000 cm^2) and 8-year old child (10,000 cm^2), and the average body weights of 70 kg for an adult and 28.2 kg for the child recommended in the EPA Exposure Factors Handbook (1996), the exposure from showering for 15 min per day with water containing different levels of arsenic have been calculated (Table 3.2). The calculations suggest that absorption through the skin surface of an adult showering with water containing 100 μg/L As(III) can account for about 20% of the recommended reference dose (0.3 μg/(kg/day) (ATSDR, 1999). The dermal route can account for over 50% of the reference dose if the shower water contains over 500 μg/L of either As(V) or As(III) for both adults and infants (Table 3.1). These calculations are consistent with epidemiological studies in Bangladesh which show that skin lesions and vascular diseases are prevalent in areas with total arsenic concentration of 100 μg/L and higher (Chowdhury et al., 1999).

Besides differences in the permeation rates, the toxicity of As(III) and As(V) also vary for different types of tissues with As(III) generally reported as being more toxic than As(V) to the skin, kidney, liver, etc (Bernstam and Nriagu, 2000a). On the other hand, As(V) is much more toxic to human red blood cells (RBC) than As(III), and this toxicity is reflected in its effects on human blood (Winski and Carter, 1998). We have demonstrated in this and other studies that the regulatory proteins of the gene expression are among the primary targets of As(III) toxicity (Bernstam and Nriagu, 2000a). This agrees with the fact that in mature RBCs, where protein synthesis does not occur and where the preferential targets for As(III) toxicity such as the signal transduction and gene expression control elements are absent, the toxicity of As(III) is lower than that of

As(V) which preferentially attacks cell membranes. Therefore, it is believed that the relative toxicity of As species may be directly related to whether the target cell is nucleated [higher As(III) toxicity] or non-nucleated as RBCs [higher toxicity of As(V)]. Apparently this phenomenon may not be unique since the differential toxic effects of sulfur-mustard have been recently related to the presence or absence of the nucleus in endothelial cells (Lodhi et al., 2001). Implicitly, one would expect the As toxicity to the hematopoietic systems to have a two-prong effect: by suppressing it at both the nucleated cell level (RBC precursors) [As(III)] as well as by destroying non-nucleated, mature RBC [As(V)]. The effects of arsenic on a given organ depends critically on the form of arsenic to which it is being exposed. Groundwater with reduced forms of arsenic will be much more toxic to the skin system compared to water in which As(V) predominates. Organ-specific toxicity of the different forms of arsenic may explain why some of the disease outcomes from exposure to arsenic (such as skin cancer, and skin hypo-pigmentation) are not uniformly observed in every part of the world.

MORPHOLOGICAL EFFECTS

The effects of absorbed arsenic on structure and integrity of the skin barrier are important issues in the dermal toxicity of this metalloid. As(III) exposure, for instance, causes skin irritation that in turn is likely to elicit the release of various cytokines capable of activating dermal phospholipases that are known to modulate skin barrier function (Mazereeuw-Hautier et al., 2000; Sjursen, Brekke, and Johansen, 2000). Exposure to As(III) may also affect the barrier lipids by modulating the activity of dermal keratinocytes and fibroblast phospholipases beyond and above the irritation and inflammation reactions that usually accompany As skin toxicity as with other xenobiotics (Schadow et al., 2001). The present studies provide evidence that artificial skins exposed to arsenate show effects on the progressive stratification and cornification of the cells that are both dose- and time-dependent (Figure 3.1). At low exposure dose of arsenate, the ultrastructural changes included development of thick and compact cornified layers as well as enlarged nuclei with margination of nuclear chromatin. Prominent keratohyalin-like granules known to contain high molecular weight proteins could be seen, indicating that an undisturbed process of terminal differentiation of keratinocytes was taking place. On the other hand, higher arsenite doses or increased exposure time resulted in a reduction in the amount of cornified cell layers, disorganization of nuclear chromatin, appearance of high number degenerated mitochondria, thickening and shortening of the tonofilaments (Figures 3.8–3.9) . At even higher doses (up to 1000 μg/L for 6 hours), we observed progressive nuclear deterioration and largely erased cellular structure, making reliable ultrastructural assessment impossible. The appearance of the submerged skin cultures after exposure to arsenic closely resembled the human skin cancers including the so-called "snake skin" appearance shown in Figure 3.2. It is not often that one is able to replicate the appearance of a diseased organ in the laboratory.

CYTOTOXICITY OF ARSENIC

Arsenite has been shown to induce cytogenetic alterations in human skin fibroblasts resulting in a significant delay of cell cycle progression, accumulation of mitotic cells, prolongation of mitosis, out-of-phase centromere separation and alterations of chromosome segregation, such as the appearance of c-metaphase, ball-metaphase, and lagged chromosomes (Yih, Ho, and Lee, 1997). Furthermore, most arsenite-treated human skin fibroblast clones manifest prolonged life span (86 ± 18 population doublings) as compared to those derived from the untreated cultures (44 ± 11 population doublings) (Yih, Ho, and Lee, 1997). Specific gene studies have demonstrated that mRNA transcription and secretion of keratinocyte growth-promoting cytokines and growth factors, including transforming growth factor-alpha (TGF alpha) and the proinflammatory cytokine tumor necrosis factor-alpha, are increased in primary human epidermal keratinocytes cultured in the presence of low concentrations of sodium arsenite (Germolec et al., 1996). At the same time, granulocyte macrophage-colony stimulating factor (GM-CSF) expression is decreased, suggesting a regulated change in specific gene expression produced by arsenic exposure resulting in cancerous changes in the skin. Furthermore, increases in AP-1 DNA binding activity, as well as c-jun and c-fos mRNA levels, occur in human keratinocytes in response to diverse dermatotoxic stimuli including arsenic as well as phorbol ester, the latter employed as a positive control (Burleson et al., 1996). The binding complex is composed of oncogenes Jun:Fos heterodimers, including Jun B and Jun D. It is important to note that since GM-CSF activates regenerative epidermal growth and stimulates keratinocyte proliferation in human skin in vivo (Braunstein et al., 1994), the effect of arsenic on this particular cytokine may be of critical importance for understanding its carcinogenic effect on skin. Detection of cytoskeletal alterations may thus provide an early indication that damage due to arsenic exposure is progressing toward malignization.

Results from the present studies show that even concentrations of As (III) below 100 μg/L produced marked changes in HK and HF morphology, and the effect on plasma membrane appeared to have caused the leakage of intracellular constituents. A significant dose-response relationship in LDH release was observed in HK exposed to As (V) and As (III), with the latter producing a 10-fold stronger LDH leakage compared with As(V). Changes in cell cytotoxicity that began with exposure to 1.0 μg/L As(III) progressively increased with concentrations of up to 1000 μg/L. In general, changes in LDH release caused by exposures to either As (V) or As(III) occurred within the first 2 hours (Figures 3.17, 3.19).

Another measure of cytotoxicity used in this project was the MTS assay which demonstrated the high susceptibility of HF to As(III) under the experimental conditions used in the study. Cytotoxicity was measurable after 48-hr exposure to 10 μg/L. Longer exposures (up to 7 days) to 10-100 μg/L As(III) produced overwhelming cytotoxicity with apparent leakage of intracellular constitutes that was manifested in a somewhat lower level of measurable formazan in exposed cells. It is possible that the growth characteristics of HF in culture might have affected their susceptibility to As(III) toxicity. The cytotoxicity tests consistently showed that the effect dose for As(III) can be as low as 1.0 μg/L, well below the current national MCL for arsenic of 10 μg/L.

Early evidence in bacterial systems indicates that arsenic specifically inhibits repair of DNA damage produced by known mutagenic agents (Rossman, Meyn, and Troll, 1975). Arsenic at concentrations of 1-5 micromole induced DNA-strand breaks associated with DNA-protein crosslinks in human fetal lung fibroblasts detectable by an alkaline elution technique and a fluorometric DNA assay (Dong and Luo, 1993). These were thought to account for chromosome aberrations and sister-chromatid exchanges induced by arsenic in vivo and in vitro. The effect of inorganic arsenicals on DNA synthesis in unsensitized human blood lymphocytes appears to be biphasic, i.e., the chemicals at very low concentrations enhanced DNA synthesis, whereas higher concentrations inhibited DNA synthesis (Meng and Meng, 1994). The stimulating effect of As(III) on DNA synthesis was stronger than As(V).

The need for studying effect of arsenic on repair of DNA damage was heightened by the report that arsenic exposure, while not producing a dramatic increase in mutations in exposed cells per se (Meng and Hsie, 1996,) was capable of enhancing the persistence of mutations in fibroblasts, thereby interfering with the repair process (Hartmann and Speit, 1996). Indeed, several lines of evidence strongly suggest that arsenic interferes with cellular repair systems and enhances mutagenicity of other agents such as UV irradiation by synergistic potentiation of the chromosomal damage produced by DNA-crosslinking agents (Yager and Wiencke, 1997). As discussed in the Introduction Section, arsenite may potentiate genetic damage through reaction with dithiols in DNA repair proteins such as poly (ADP-ribose) polymerase (PARP), perhaps resulting in interference with normal repair function (Yager and Wiencke, 1997).

Limited assessments of DNA repair/damage produced by arsenic have been made for arsenic-treated skin system, particularly at concentrations found in drinking water. This report provides some insight into the effects of arsenic exposure on DNA, RNA and protein synthesis using the incorporation of ^{3}H-labeled compounds. Since it appears that at least some of the malignant potential of arsenic exposure lies in the disruption of the repair ability of exposed cells, this study has included the time course of recovery of DNA, RNA and protein synthesis affected by arsenic exposure. The results are consistent with the report by Germolec et al. (1996) in that treatment of human keratinocytes with micromolar concentrations of sodium arsenite resulted in a significant increase in cell proliferation as indicated by increases in cell numbers, c-myc gene expression and incorporation of ^{3}H-thymidine into cellular DNA. The normal kinetics of DNA replication and mitosis in submerged cultures of HK were defined for the first 7 days after seeding (Ku and Bernstein, 1988). Essentially all cells underwent replication of DNA followed by mitosis within the first 3 days in culture, while only about 60% of the cells were proliferative by 6 days in culture. As noted previously, lifted human keratinocyte cultures showed significant alterations in the morphology of the nuclei stemming from clumping and margination of chromatin. All of those results are best rationalized as indicative of damage in the DNA repair (Figures 3.21–3.23) and protein synthesis (Figure 3.26) apparatus in the exposed cells.

Higher toxicity of As(III) compared to As(V) was demonstrated in relation to the DNA integrity, as represented by in situ DNA fragmentation at the single cell level using the TdT-mediated biotin dUTP nick end-labelling (TUNEL assay) (Bernstam and Nriagu, 2000b). Dose- and time-dependent relationships were found between exposure to relatively low levels (5-100 μg/L) of As(III) and single strand DNA breaks and alkali labile damage in HK and in HF (Figure 3.32). Programmed cell death—apoptosis—which was also observed at low doses of human skin cells to As(III), presumably involved a coordinated engagement of multiple signal transduction cascades leading to DNA fragmentation and eventual cell death.

Another approach for the identification of cytotoxic effects of xenobiotics is the single cell gel electrophoresis assay ("comet assay") that has been shown to be a much more sensitive technique than micronuclei assay for detection of DNA strand breaks produced by exposure to low levels of toxicants (Tates et al., 1996; Van Goethem, Lison, and Kirsch-Volders, 1997; Villani et al., 1998). The comet assay has been widely used to monitor effects of human carcinogen on the DNA (for a review see: Anderson, Yu, and McGregor, 1998) as well as to evaluate genotoxic consequences of chronic low level exposures to radiation (Plappert et al., 1997). The comet assay can also differentiate between cytotoxins and genotoxins on the basis of the comets profiles derived from treated cells (Henderson et al., 1998). Comparison between the comet patterns obtained in regular and alkaline versions of the assay can reveal how a toxic agent affects DNA supercoiling versus its fragmentation (Duthie et al., 1997). Since the comets that extend from cells embedded in agarose on a slide are visualized and measured on an individual cell basis, the comet assay offers the additional advantage over DNA unwinding assays by allowing easy estimates of heterogeneity in the cell response (Olive et al., 1997). The pattern of repair of the DNA strand break damage may offer important insights into the mechanisms of cellular resistance and adaptation to low level chronic exposure to arsenic in the environment.

This study found dose- and time-related changes in the comet moment of skin cells exposed to As(III) and As(V) concentrations as low as 10 μg/L (Figures 3.30–3.33). The results were consistent with the effects of As on apoptotic changes obtained by the TUNEL assay. It appeared that at low concentrations of As(III), mitochondrial damage somewhat exceeded that observed in nuclei of basal cells, but additional work is needed to confirm this observation. It was not clear whether the injuries reflected an erroneous DNA repair or were a direct effect of the As(III) on the cellular structures. It would appear that molecular damage to DNA determined by the TUNEL and COMET assays preceded detectable ultrastructural changes in the cells studied.

GENOTOXICITY OF ARSENIC

Arsenic, a unique human carcinogen, is a potent clastogen in a variety of mammalian cell systems. Arsenite induces many types of cytogenetic alterations, such as hyperploidy, breakage, sister chromatid exchanges, chromosome aberrations, and endoreduplication in a variety of in vivo and in vitro systems (Rupa, Schuler, and Eastmond, 1997). Cytogenetic alterations are frequently associated with cancer development. Alterations in chromosome segregation have been observed in individuals exposed to high concentrations of arsenic in the drinking water (Vega, Gonsebatt, and Ostrosky-Wegman, 1995), and a dose-related effect of arsenic on chromosome aberrations produces a colchicine-like effect (Vega, Gonsebatt, and Ostrosky-Wegman, 1995). Arsenic exposure has been repeatedly shown to produce chromosome aberrations in human cells (Gurr et al., 1998; Huang et al., 1995; Kochhar et al., 1996). The effect of sodium arsenite on chromatid separation in human skin fibroblasts results in arrest of cells in the G2 phase, interference with mitotic division, inhibition of spindle assembly, and induction of chromosome endoreduplication in their second mitosis (Huang et al., 1995).

Telomeres are chromosome structures that provide stability to chromosome ends, and the length of telomeres shortens with progressive cell divisions. Loss of telomers may result in genomic instability, abnormal chromosomes, and plays a role in cellular senescence (Shay, 1995; Taylor et al., 1996). Telomerase, a novel ribonucleoprotein, adds nucleotides to the ends of chromosomes during DNA replication. Telomerase activity has been found in malignant and nonmalignant skin conditions: skin damaged by UV radiation, hyperproliferative psoriasis, and poison ivy dermatitis (Taylor et al., 1996). On the other hand, chromosomal damage by arsenic exposure has been well-established (e.g., Burgdorf, Kurvink, and Cervenka, 1977; McCabe, McGuire, and Nowak, 1983; Vig et al., 1984; Gonsebatt et al., 1994; Lerda, 1994; Huang et al., 1995). Intuitively, arsenic should be able to induce telomerase activity, which was exactly demonstrated, for the first time, in this study. Exposure of HF to As(III) for a short period of time (15 min) was able to increase the telomerase activity by some 37-67% over that observed in unexposed control cells (Figure 3.34). Lengthening of exposure time resulted in significant changes in telomerase activity in cultured fibroblast, suggesting that an increase in exposure period may be an important factor that determines carcinogenicity of arsenic.

Arsenic is capable of inducing heat shock proteins (HSPs) of various sizes, from the smaller proteins such as alphaB-crystallin, HSP25, HSP27, HSP30, etc (Huot et al., 1995; Haynes, Duncan, and Piatigorsky, 1996; Ito et al., 1997; Zavialov et al., 1998; Lambert et al., 1999) to the larger HSP species such as HSP105 (Yasuda et al., 1995). While some HSPs are constituitively expressed, others are only induced by various stress agents. In their general features, HSPs are remarkably similar in different animal, bacterial and plant species (for a comprehensive review see Feder and Hofmann, 1999). However, specifics of activation or suppression of the signal transduction pathway may be highly selective with respect to cell type, stress agent, its dose and metabolic state of the cell. Activation of the heat shock factor (HSF) leads to an increase in the HSP70 family gene expression (Gomer et al., 1996; Lee and Corry, 1998), for instance. On the other hand, overexpression of HSP70 inhibits the activation of its transcriptional factor HSF-1 (Ding, Tsokos, and Kiang, 1998). HSF-1, in turn, inhibits the expression of c-fos, an immediate early gene that controls responses to extracellular stimuli for growth and differentiation (Chen et al., 1997). Through the cascade of changes in stress-related signal transduction pathways and the involvement of transcription regulators, the genotoxic effects of arsenic stress can activate or suppress gene expression, eventually resulting either in cell adaptation, malignization or death.

This project has explored the effects of As(III) in enhancing the expression of selected stress-related proteins in the signal transduction pathway. It found that HK exposed to 10–500 μg/L of As(III) displayed a dose-dependent increase in HSP70 expression detected by Western Blot technique (Figure 3.36A). When estimating the extent of the 7-fold increase in HSP70 expression observed, the progressive loss of protein in HK exposed to these As(III) concentrations should be taken into account.

A similar dose-dependent increase in the expression of the signal transduction protein HSP38 was observed in HK exposed to 5-500 μg/L of As(III) (Figure 3.36B). Mutations in the protein 53 (p53) gene are often found in human cancer (Levin, 1997). An important function of p53 is tumor suppression by counteracting the carcinogenic influences largely by inducing apoptosis. It has been established that p53 performs this protective function also when cells are exposed to wide range of stressors, including exposure to xenobiotics. In normal cells, the detectable expression of p53 is very low, mostly because of the short half-life of the protein, and can hardly be detected using conventional immunohistological techniques. In response to stress signals (DNA damage, hypoxia, etc.) the p53 can accumulate in the organs (Lowe et al., 1993). Therefore, the presence of p53 expression reflects a significant activation of expression of normal wild type p53. This reflects the cells' effort to counteract potentially genotoxic stress.

Results of studies conducted with different technique converge to show that exposure to As(III) concentration of 100 μg/L can cause serious metabolic disturbances in human skin cells leading to the formation of p53 (Figure 3.37). The consequences of such disturbances in terms of long-term development of gross clinical abnormalities or skin pathology are unknown.

CHEMICAL INHIBITION OF ARSENIC TOXICITY

Potential inhibition of arsenic toxicity by chemical agents is a topic that has not received much attention. Our cultured human skin system provides a unique vehicle for this type of study. We investigated the effect of arsenic on HK in the presence of several MAPK inhibitors. These inhibitors interefere with the triggering of a cascade of cellular events by a variety of stressors (including toxic metals) that may end in apoptosis through the activation of intracellular signaling pathways. Interestingly, the inhibition experiments served to probe the mechanisms of arsenic toxicity. Our initial working hypothesis was that similar to inflammation where inhibition of prostaglandin synthesis alleviates the damaging consequences of exposure to radiation or toxic agents (Leong et al., 1996; Muller-Decker et al., 1998; Rys-Sikora et al., 2000; Brouxhon et al., 2001), the use of specific inhibitors that disrupt the prostaglandin synthesis (COX-1 and COX-2 inhibitors) could significantly reduce the arsenite-related damage.

The protective effects of MAPK inhibitors, namely, SB 203580 and SB 202190, against As(III) toxicity are similar to the known inhibition of the expression of heat shock proteins induced by various cellular stressors including toxic metals. This provides clear evidence for the similarity between the basic features of As(III) toxicity and the effects of heat stress (Bernstam and Nriagu, 2000b). The results obtained point to high specificity of SB 203580 and SB 202190 for the inhibition of activation of p38 by phosphorylation. With exposure to the lower As(III) doses (5-10 μg/L), the activation of p38 was primarily in the cytoplasm, which is characteristic of the initial stages of activation of the signal transduction pathways. With progressive toxicity as seen at higher doses of As(III), the activated p38 is more associated with the cell nucleus. In other words, the p38 activation is largely limited to the nucleus in HF cells exposed to the high As(III) concentrations of 500 and 1000 μg/L (Figure 3.43D). The high deterioration of cells exposed to 1000 μg/L prevented successful micrographs to be taken.

The protective effect of MAPK inhibitors leading to the preservation of morphological integrity of cells exposed to As(III) in all probability is a multifactorial event in which inhibitor's ability to interrupt COX-2 biosynthesis may have played an important role. This, however, can not constitute the main or the only route of protective action by the MAPK inhibitors since blocking of COX-2 activity by specific inhibitors failed to exert strong protective action against As(III) toxicity. No measurement of COX-2 activity was made in this study and the efficiency of COX-2 inhibitors and their concentrations used in this study will have to be independently established before any conclusive evidence for the protective role of COX-2 inhibition against As(III) toxicity can be drawn.

CHAPTER 5
RECOMMENDATIONS

This project is, to a large extent, exploratory and was done in context of rapidly expanding knowledge on arsenic toxicity at the cellular, systemic, and community levels. The focus of the study has been on understanding the processes and mechanisms of arsenic toxicity at the molecular level, and there is clearly a need to apply the knowledge to the study of systemic toxicity of arsenic. For instance, the transfer and retention of arsenic in different layers of the skin was not addressed, nor was the metabolism of arsenic in the various skin components evaluated in any meaningful way. The use of artificial human skin cultures precluded any study of exchange of arsenic between the blood stream and the skin.

An aspect of arsenic toxicity that has not received due attention is the apparent ability to cause cooperative phase transitions in cell membrane properties as well as in DNA synthesis. While the resemblance of the effects of arsenic toxicity to those of heat shock on signal transduction pathways and other metabolic systems in the cell has been generally known, there has been no previous work on the ability of arsenic to produce similar effects in signal transduction. The pattern of transition of the resistant state to loss of selective permeability in human keratinocytes exposed to low doses of arsenic strongly suggests that such cooperative transitions do take place.

More experimental work needs to be done to establish the details of arsenic-related phase transitions in cell membranes and other cellular systems such as DNA synthesis, which can lead to a better understanding of the mechanisms of arsenic toxicity. Epidemiological risks of long term human exposure to low levels of arsenic that may result in various pathological conditions can be better understood in the context of the cumulative effects of low doses of arsenic on selective cell permeability that abruptly fails leading to a breakdown in skin barrier function and engendering unimpeded access of arsenic into the cells. Various environmental factors (temperature, sweat, humidity, etc) can affect the degree of the cooperative phase transition at a given arsenic exposure dose, and it would be interesting to evaluate their influence on the skin barrier function and macromolecular properties when the resistance threshold is exceeded from to exposure to arsenic.

Recent evidence suggests that alterations in tumor suppressor gene function are responsible for malignization of cells. Different concentrations of sodium arsenite have been found to produce dose-related inhibition of mitotic response and lymphocyte proliferation and to induce the expression of tumor suppressor gene p53 in different cell lines (HeLa, C-33) and a lymphoblast cell line transformed with Epstein-Barr virus (LCL-EBV) (Salazar et al., 1997). These reports indicate that arsenic induces p53 expression and that cells with a functional p53 content would better deal with damage induced by this metalloid (Salazar et al., 1997). While lacking direct mutagenic activity, arsenic may act as a tumor promoter by stimulating the transcription factor AP-1 and efficiently inducing c-fos and c-jun gene expression similar to other tumor promoters (Cavigelli et al., 1996). Activation of AP-1 helps regulate the expression of genes involved in cell growth, inflammatory responses, and repair processes. In turn, AP-1 binding is associated with an increase in IL-1 alpha expression, an early mediator of toxic response in the skin (Burleson et al., 1996). While significant changes in cytokine gene expression have been observed in animal skin carcinogenesis, no p53 inactivation has been reported (Shibata et al., 1997). Since there are significant differences in sensitivity of specific genes in animals and humans, it would be interesting to undertake additional studies to evaluate (i) the mechanistic role of arsenic in human cancers,

(ii) the earliest stages of signal transduction and the critical events in transcription regulation with emphasis on revealing similarities and distinctions with the effects of heat stress that displays significant parallels to As toxicity, (iii) effects on regulators of the inflammatory cascade including the prostaglandin synthesis and regulation of cox-1 and cox-2 expression, and (iv) the effectiveness of specific inhibitors of the signal transduction pathways in preventing the development of irreversible changes in As-exposed cells. An assessment of the conditions and specific agents promoting the repair of protein and DNA syntheses especially during chronic, long-term exposure to arsenic will be an important adjunct to the these cytotoxic and genotoxic studies.

Since arsenic is believed to be detoxified via methylation using a methyltransferase (MTase) and S-adenosylmethionine as the methyl donor, suggestions have been made that arsenic carcinogenesis at the DNA level may result from hypomethylation of DNA due to competition for methyl donating group with arsenic (Zhao et al., 1997; Tice et al., 1997). Arsenic-induced DNA hypomethylation was found to be a function of dose and exposure duration and remained constant even after withdrawal of arsenic exposure. Hyperexpressibility of the MT gene, which is controlled by DNA methylation, was also detected in transformed cells (Zhao et al., 1997). Interestingly, acute arsenic or arsenic at nontransforming levels did not induce global hypomethylation of DNA. In fact, it has been reported in cultured human lung cells that arsenic alters cytosine methylation patterns of the promoter of the tumor suppressor gene p53, suggesting a model of arsenic carcinogenicity related to alteration of DNA methylation by arsenic (Mass and Wang, 1997). There is a need for detailed investigation of methylation/demethylation of arsenic in cells to improve understanding this phenomenon and its role in the toxicity of arsenic.

The preliminary results obtained suggest that the multifactorial effects of MAKP inhibition of As(III) toxicity, even at the highest doses of As(III) tested, are mediated through a variety of pathways including transcriptional effects. Additional research should be directed at the analysis of the arsenic-induced specific events in signal transduction pathways that are being interrupted by these inhibitors. An important issue to be resolved would be to establish the earliest point in time of the cell commitment to the stress response and the lowest effective inhibitor concentrations that can be used to interrupt this destructive process. Conceivably, such studies can lead to the development of chemical protective measures against the toxic effects of environmental or occupational exposure to arsenic.

REFERENCES

Abe, K., and Matsuki, N. (2000.) Measurement of cellular 3-(4,5-dimethylthiazol-2-yl)-2,5-diphenyltetrazolium bromide (MTT) reduction activity and lactate dehydrogenase release using MTT. *Neurosci. Res.*, 38: 325-329.

Abello, P.A., and Buchman, T.G. (1994.) Heat shock-induced cell death in murine microvascular endothelial cells depends on priming with tumor necrosis factor-alpha or interferon-gamma. *Shock*, 2: 320-323.

Abello, P.A.; Fidler, S.A.; and Buchman, T.G. (1994.) Thiol reducing agents modulate induced apoptosis in porcine endothelial cells. *Shock*, 2: 79-83.

Ahmad, A.; Bondaranayake, D.; Khan, A.W.; Hadi, A.; Uddin, G.; and Halim, A. (1997.) Arsenic contamination in groundwater and arsenicosis in Bangladesh. *Int. J. Environ. Health Res.*, 7: 271-276.

Akao, Y.; Mizoguchi, H.; Kojima, S.; Naoe, T.; Ohishi, N.; and Yagi, K. (1998.) Arsenic induces apoptosis in B-cell leukaemic cell lines in vitro: Activation of caspases and down-regulation of Bcl-2 protein. *Br. J. Haematol.*, 102: 1055-1060.

Albores, A.; Cebrian, M.E.; Garcia-Vargas, G.G.; Connelly, J.C.; Price, S.C.; Hinton, R.H.; Bach, P.H.; and Bridges, J.W. (1996.) Enhanced arsenite-induced hepatic morphological and biochemical changes in phenobarbital-pretreated rats. *Toxicol. Pathol.*, 24: 172-180.

Ali, A.; Krone, P.H.; Pearson, D.S.; and Heikkila, J.J. (1996.) Evaluation of stress-inducible hsp90 gene expression as a potential molecular biomarker in *Xenopus laevis*. *Cell Stress Chaperones*, 1: 62-69.

Amici, C.; Giorgi, C.; Rossi, A.; and Santoro, M.G. (1994.) Selective inhibition of virus protein synthesis by prostaglandin A1: A translational block associated with HSP70 synthesis. *J. Virol.*, 68: 6890-6899.

Anderson, D.; Yu, T.W.; and McGregor, D.B. (1998.) Comet assay responses as indicators of carcinogen exposure. *Mutagenesis*, 13: 539-55.

Aposhian, H.V.; Zakharyan, R.; Wu, Y.; Healy, S.; and Aposhian, M.M. (1997.) Enzymatic methylation of arsenic compounds. In *Arsenic Exposure and Health Effects*. Edited by C.O. Abernathy, R.L. Calderon, and W.R. Chappell. New York: Chapman & Hall. pp. 296-321.

ATSDR. (1999.) *Toxicological Profile for Arsenic*. Agency for Toxic Substances and Disease Registry, Public Health Service, Agency for Toxic Substances and Disease Registry.

Bates, M.N.; Smith, A.H.; and Hopenhayn-Rich, C. (1992.) Arsenic ingestion and internal cancers: A review. *Am. J. Epidemiol.*, 135: 462-476.

Bazarbachi, A.; El-Sabban, M.E.; Nasr, R.; Quignon, F.; Awaraji, C.; Kersual, J.; Dianoux, L.; Zermati, Y.; Haidar, J.H.; de Hermine, O.H. (1999.) Arsenic trioxide and interferon-alpha synergize to induce cell cycle arrest and apoptosis in human T-cell lymphotropic virus type I-transformed cells. *Blood*, 93: 278-283.

Beaubien, S.E.; Nriagu, J.O.; Lawson, G.; and Blowes, D.W. (1993.) Speciation and distribution of chromium in the Great Lakes. *Environ. Sci. Technol.*, 26: 730-736.

Bernstam, L.; Lan, C.H.; Lee, J.; and Nriagu, J.O. (2002.) Effects of arsenic on human keratinocytes. Morphological, physiological and precursor incorporation studies. *Environ. Res.*, 89: 220-235.

Bernstam, L., and Nriagu, J. (2000a.) Apoptotic and signal transduction changes in human keratinocytes exposed to As(III) and As(V). *11th Annual Intl. Conf. on Heavy Metals in the Environment.* Contr. #1026. Ann Arbor, Mich.

Bernstam, L., and Nriagu, J. (2000b.) Molecular aspects of arsenic stress. *J. Toxicol. Environ. Health*, Part B, 3: 293-322.

Bernstam, L.; Vaughan, F.L.; and Bernstein, I.A. (1990.) Stratified cornified primary cultures of human keratinocytes grown on microporous membranes at the air-liquid interface. *J. Dermatol. Sci.*, 1: 173-181.

Bernstam, L.L.; Vaughan, F.L.; and Bernstein, I.A. (1986.) Keratinocytes grown at the air-liquid interface. In Vitro Devel. *Cell Biol.*, 22: 695-703.

Bittel, D.; Dalton, T.; Samson, S.L.-A.; Gedamu, L.; and Andrews, G.K. (1998.) The DNA binding activity of metal response element-binding transcription factor-1 is activated in vivo and in vitro by zinc, but not by other transition metals. *J. Biol. Chem.*, 273: 7127-7133.

Boyce, S., and Ham, R. (1983.) Normal human epidermal keratinocytes. In *In Vitro Models for Cancer Research*, Vol. III. Edited by M.M. Weber and L. Sekely. Boca Raton, Fla.: CRC Press. pp. 245-274.

Braunstein, S.; Kaplan, G.; Gottlieb, A.B.; Schwartz, M.; Walsh, G.; Abalos, R.M.; Fajardo, T.T.; Guido, L.S.; and Krueger, J.G. (1994.) GM-CSF activates regenerative epidermal growth and stimulates keratinocyte proliferation in human skin in vivo. *J. Invest. Dermatol.*, 103: 601-4.

Brostrom, C.O., and Brostrom, M.A. (1998.) Regulation of translational initiation during cellular responses to stress. *Prog. Nucleic Acid Res. Mol. Biol.*, 58: 79-125.

Brostrom, C.O.; Prostko, C.R.; Kaufman, R.J. and Brostrom, M.A. (1996.) Inhibition of translational initiation by activators of the glucose-regulated stress protein and heat shock protein stress response systems. Role of the interferon-inducible double-stranded RNA-activated eukaryotic initiation factor 2alpha kinase. *J. Biol. Chem.*, 271: 24995-25002.

Brouxhon, S.M.; Schneider, S.; Shenkut, E.; and Pentland, A. (2001.) The role of cyclooxygenase-2 in skin burn injury. *Acad. Emerg. Med.*, 8: 520-521.

Brown, R.; Ku, W.; and Bernstein, I.A. (1987.) Changes in lectin binding by differentiating cutaneous keratinocytes from the newborn rat. *J. Invest. Dermatol.*, 88: 719-726.

Buchet, J.P., and Lauwerys, R. (1985.) Study of inorganic arsenic methylation by rat in vitro: Relevance for the interpretation of observations in man. *Arch. Toxicol.*, 57: 125-129.

Burgdorf, W.; Kurvink, K.; and Cervenka, J. (1977.) Elevated sister chromatid exchange rate in lymphocytes of subjects treated with arsenic. *Hum. Genet.*, 36: 69-72.

Burleson, F.G.; Simeonova, P.P.; Germolec, D.R.; and Luster, M.I. (1996.) Dermatotoxic chemical stimulate of c-jun and c-fos transcription and AP-1 DNA binding in human keratinocytes. *Res. Commun. Mol. Pathol. Pharmacol.*, 93: 131-48.

Cai, J.; Salmon, K.; and DuBow, M.S. (1998.) A chromosomal ars operon homologue of *Pseudomonas aeruginosa* confers increased resistance to arsenic and antimony in *Escherichia coli. Microbiol.*, 144 (Pt 10): 2705-2713.

Cavigelli, M.; Li, W.W.; Lin, A.; Su, B.; Yoshioka, K.; and Karin, M. (1996.) The tumor promoter arsenite stimulates AP-1 activity by inhibiting a JNK phosphatase. *EMBO J.*, 15: 6269-6279.

Chai, C.Y.; Yu, H.S.; Yen, H.T.; Tsai, K.B.; Chen, G.S.; and Yu, C.L. (1997.) The inhibitory effect of UVB irradiation on the expression of p53 and Ki-67 proteins in arsenic-induced Bowen's disease. *J. Cutan. Pathol.*, 24: 8-13.

Chen, C.; Xie, Y.; Stevenson, M.A.; Auron, P.E.; and Calderwood, S.K. (1997.) Heat shock factor 1 represses ras-induced transcriptional activation of the *c-fos* gene. *J. Biol. Chem.*, 272: 26803-26806.

Chen, C.J.; Chuang, Y.C.; Lin, T.M.; and Wu, H. (1985.) Malignant neoplasms among residents of a Blackfoot disease-endemic area in Taiwan: High arsenic artesian well water and cancers. *Cancer Res.*, 45: 5895-5988.

Chen, G.Q.; Zhu, J.; Shi, X.G.; Ni, J.H.; Zhong, H.J.; Si, G.Y.; Jin, X.L.; Tang, W.; Li, X.S.; Xong, S.M.; Shen, Z.X.; Sun, G.L.; Ma, J.; Zhang, P.; Zhang, T.D.; Gazin, C.; Naoe, T.; Chen, S.J.; Wang, Z.Y.; and Chen, Z. (1996.) In vitro studies on cellular and molecular mechanisms of arsenic trioxide in the treatment of acute promyelocytic leukemia. *Blood*, 88(3): 1052-1061.

Chen, Y.C.; Guo, Y.L.; Su, H.J.; Hsueh, Y.M.; Smith, T.J.; Ryan, L.M.; Lee, M.S.; Chao, S.C.; Lee, J.Y.Y.; and Christiani, D.C. (2003.) Arsenic methylation and skin cancer risk in southwestern Taiwan. *J. Occup. Environ. Med.*, 45: 241-248.

Choudhury, A.R.; Das, T.; and Sharma, A. (1997.) Mustard oil and garlic extract as inhibitors of sodium arsenite-induced chromosomal breaks in vivo. *Cancer Lett.*, 121: 45-52.

Choudhury, A.R.; Das, T.; Sharma, A.; and Talukder, G. (1997.) Inhibition of clastogenic effects of arsenic through continued oral administration of garlic extract in mice in vivo. *Mutat. Res.*, 392: 237-242.

Chowdhury, U.K.; Biswas, B.K.; Dhar, R.K.; Samanta, G.; Mandal, B.K.; Chowdhury, T.R.; Chakraborti, D.; Kabir, S.; and Roy, S. (1999.) Groundwater arsenic contamination and suffering of people in Bangladesh. In *Arsenic Exposure and Health Effects.* Edited by W.R. Chappell, C.O. Abernathy, and R.L. Calderon. Amsterdam: Elsevier. pp. 165-182.

Chu, W.A.; Moehlenkamp, J.D.; Bittel, D.; Andrews, G.K.; and Johnson, J.A. (1999.) Cadmium-mediated activation of the metal response element in human neuroblastoma cells lacking functional metal response element-binding transcription factor-1. *J. Biol. Chem.*, 274: 5279-5284.

Corton, J.M.; Gillespie, J.G.; and Hardie, D.G. (1994.) Role of the AMP-activated protein kinase in the cellular stress response. *Curr. Biol.*, 4: 315-324.

Cory, A.H.; Owen, T.C.; Barltrop, J.A.; and Cory, J.G. (1991.) Use of an aqueous soluble tetrasolium/formazan assay for cell growth assays in culture. *Cancer Commun.*, 3: 207-212.

Cowlishaw, J.L.; Pollard, E.J.; Cowen, A.E.; and Powell, L.W. (1979.) Liver disease associated with chronic arsenic ingestion. *Aust. N. Z. J. Med.* 9: 310-313.

Crameri, A.; Dawes, G.; Rodriguez Jr., E.; Silver, S.; and Stemmer, W.P. (1997.) Molecular evolution of an arsenate detoxification pathway by DNA shuffling. *Nat. Biotechnol.*, 15: 436-438.

Cristofalo, V.J.; Volker, C.; Francis, M.K.; and Tresini, M. (1998.) Age-dependent modifications of gene expression in human fibroblasts. *Crit. Rev. Eukaryot. Gene Expr.*, 8: 43-80.

Cullen, K.E., and Sarge, K.D. (1997.) Characterization of hypothermia-induced cellular stress response in mouse tissues. *J. Biol. Chem.*, 272: 1742-1746.

Decker, T., and Lohmann-Matthes, M.-L. (1988.) A quick and simple method for the quantitation of lactate dehydrogenase release in measurements of cellular cytotoxicity and tumor necrosis factor (TNF) activity. *J. Immunol. Methods*, 15: 61-69.

DeMeester, S.L.; Buchman, T.G.; Qiu, Y.; Dunnigan, K.; Hotchkiss, R.S.; Karl, I.E.; and Cobb, J.P. (1997.) Heat shock induces IkappaB-alpha and prevents stress-induced endothelial cell apoptosis. *Arch. Surg.*, 132: 1283-1288.

DeMeester, S.L.; Buchman, T.G.; Qiu, Y.; Dunnigan, K.; Hotchkiss, R.S.; Karl, I.E.; and Cobb, J.P. (1998a.) Pyrrolidine dithiocarbamate activates the heat shock response and thereby induces apoptosis in primed endothelial cells. *Shock*, 10: 1-6.

DeMeester, S.L.; Cobb, J.P.; Hotchkiss, R.S.; Osborne, D.F.; Karl, I.E.; Tinsley, K.W.; and Buchman, T.G. (1998b.) Stress-induced fractal rearrangement of the endothelial cell cytoskeleton causes apoptosis. *Surgery*, 124: 362-371.

de Vera, M.E.; Kim, Y.M.; Wong, H.R.; Wang, Q.; Billiar, T.R.; and Geller, D.A. (1996a.) Heat shock response inhibits cytokine-inducible nitric oxide synthase expression in rat hepatocytes. *Hepatology*, 24: 1238-1245.

de Vera, M.E.; Wong, J.M.; Zhou, J.Y.; Tzeng, E.; Wong, H.R.; Billiar, T.R.; and Geller, D.A. (1996b.) Cytokine-induced nitric oxide synthase gene transcription is blocked by the heat shock response in human liver cells. *Surgery*, 120: 144-149.

Dewji, N.N., and Do, C. (1996.) Heat shock factor-1 mediates the transcriptional activation of Alzheimers beta-amyloid precursor protein gene in response to stress. *Brain Res. Mol. Brain Res.*, 35: 325-328.

Ding, X.Z.; Tsokos, G.C.; and Kiang, J.G. (1998.) Overexpression of HSP-70 inhibits the phosphorylation of HSF1 by activating protein phosphatase and inhibiting protein kinase C activity. *FASEB J.*, 12: 451-459.

Dix, D.J.; Garges, J.B.; and Hong, R.L. (1998.) Inhibition of hsp70-1 and hsp70-3 expression disrupts preimplantation embryogenesis and heightens embryo sensitivity to arsenic. *Mol. Reprod. Dev.*, 51: 373-380.

Dong, J.T., and Luo, X.M. (1993.) Arsenic-induced DNA-strand breaks associated with DNA-protein crosslinks in human fetal lung fibroblasts. *Mutat. Res.*, 302(2): 97-102.

Dubois, M.F.; Bellier, S.; Seo, S.J.; and Bensaude, O. (1994.) Phosphorylation of the RNA polymerase II largest subunit during heat shock and inhibition of transcription in HeLa cells. *J. Cell. Physiol.*, 158: 417-426.

Duthie, S.J.; Collins, A.R.; Duthie, G.G.; and Dobson, V.L. (1997.) Quercetin and myricetin protect against hydrogen peroxide-induced DNA damage (strand breaks and oxidised pyrimidines) in human lymphocytes. *Mutat. Res.*, 393(3): 223-31.

Dutkiewicz, T. (1977.) Experimental studies on arsenic absorption routes in rats. *Environ. Health Perspect.*, 19: 173-177.

Eguchi, N.; Kuroda, K.; and Endo, G. (1997.) Metabolites of arsenic induced tetraploids and mitotic arrest in cultured cells. *Arch. Environ. Contam. Toxicol.*, 32: 141-145.

Elbirt, K.K.; Whitmarsh, A.J.; Davis, R.J.; and Bonkovsky, H.L. (1998.) Mechanism of sodium arsenite-mediated induction of heme oxygenase-1 in hepatoma cells. Role of mitogen-activated protein kinases. *J. Biol. Chem.*, 273: 8922-8931.

Elia, G.; De Marco, A.; Rossi, A.; and Santoro, M.G. (1996.) Inhibition of HSP70 expression by calcium ionophore A23187 in human cells. An effect independent of the acquisition of DNA-binding activity by the heat shock transcription factor. *J. Biol. Chem.*, 271: 16111-16118.

Everall, J.D., and Dowd, P.M. (1978.) Influence of environmental factors excluding ultra violet radiation on the incidence of skin cancer. *Bull. Cancer*, 65: 241-247.

Feder, M.E., and Hofmann, G.E. (1999.) Heat-shock proteins, molecular chaperones and the stress response: Evolutionary and ecological physiology. *Annu. Rev. Physiol.*, 61: 243-282.

Gabai, V.L.; Meriin, A.B.; Mosser, D.D.; Caron, A.W.; Rits, S.; Shifrin, V.I.; and Sherman, M.Y. (1997.) Hsp70 prevents activation of stress kinases. *J. Biol. Chem.*, 272: 18033-18037.

Gebel, T.; Birkenkamp, P.; Luthin, S.; and Dunkelberg, H. (1998.) Arsenic(III), but not antimony(III), induces DNA-protein crosslinks. *Anticancer Res.*, 18: 4253-4257.

Germolec, D.R.; Yoshida, T.; Gaido, K.; Wilmer, J.L.; Simeonova, P.P.; Kayama, F.; Burleson, F.; Dong, W.; Lange, R.W.; and Luster, M.I. (1996.) Arsenic induces overexpression of growth factors in human keratinocytes. *Toxicol. Appl. Pharmacol.*, 141: 308-318.

Ghoshal, K.; Wang, Y.; Sheridan, J.F.; and Jacob, S.T. (1998.) Metallothionein induction in response to restraint stress. Transcriptional control, adaptation to stress, and role of glucocorticoid. *J. Biol. Chem.*, 273: 27904-27910.

Goering, P.L.; Aposhian, H.V.; Mass, M.J.; Cebrian, M.; Beck, B.D.; and Waalkes, M.P. (1999.) The enigma of arsenic carcinogenesis: Role of metabolism. *Toxicol. Sci.*, 49: 5-14.

Gomer, C.J.; Ryter, S.W.; Ferrario, A.; Rucker, N.; Wong, S.; and Fisher, A.M. (1996.) Photodynamic therapy-mediated oxidative stress can induce expression of heat shock proteins. *Cancer Res.*, 56: 2355-2360.

Gonsebatt, M.E.; Vega, L.; Montero, R.; Garcia-Vargas, G.; Del Razo, L.M.; Albores, A.; Cebrian, M.E.; and Ostrosky-Wegman, P. (1994.) Lymphocyte replicating ability in individuals exposed to arsenic via drinking water. *Mutat. Res.*, 313: 293-299.

Gordon, S.A.; Hoffman, R.A.; Simmons, R.L.; and Ford, H.R. (1997.) Induction of heat shock protein 70 protects thymocytes against radiation-induced apoptosis. *Arch. Surg.*, 132: 1277-1282.

Gurr, J.R.; Liu, F.; Lynn, S.; and Jan, K.Y. (1998.) Calcium-dependent nitric oxide production is involved in arsenite-induced micronuclei. *Mutat. Res.*, 416(3): 137-148.

Guy, R.H.; Hostynek, J.J.; Hinz, R.S.; and Lorence, C.R. (1999.) *Metals and the Skin*. New York: Marcel Dekker. pp. 63-74.

Hartmann, A., and Speit, G. (1996.) Effect of arsenic and cadmium on the persistence of mutagen-induced DNA lesions in human cells. *Environ. Mol. Mutagen.*, 27: 98-104.

Hatayama, T.; Takahashi, H.; and Yamagishi, N. (1997.) Reduced induction of HSP70 in PC12 cells during neuronal differentiation. *J. Biochem. (Tokyo)*, 122: 904-910.

Hauser, G.J.; Dayao, E.K.; Wasserloos, K.; Pitt, B.R.; and Wong, H.R. (1996.) HSP induction inhibits iNOS mRNA expression and attenuates hypotension in endotoxin-challenged rats. *Am. J. Physiol.*, 271: H2529-H2535.

Haynes II, J.I.; Duncan, M.K.; and Piatigorsky, J. (1996.) Spatial and temporal activity of the alpha B-crystallin/small heat shock protein gene promoter in transgenic mice. *Dev. Dyn.*, 207: 75-88.

Henderson, L.; Wolfreys, A.; Fedyk, J.; Bourner, C.; and Windebank, S. (1998.) The ability of the Comet assay to discriminate between genotoxins and cytotoxins. *Mutagenesis*, 13: 89-94.

Hennings, H.; Michael, D.; Cheng, C.; Steinert, P.; Holbrook, K.; and Yuspa, S.H. (1980.) Calcium regulation of growth and differentiation of mouse epidermal cells in culture. *Cell*, 19: 245-254.

Henriksson, J.; Johannisson, A.; Bergqvist, P.A.; and Norrgren, L. (1996.) The toxicity of organoarsenic-based warfare agents: in vitro and in vivo studies. *Arch. Environ. Contam. Toxicol.*, 30: 213-219.

Hopenhayn-Rich, C.; Biggs, M.L.; Fuchs, A.; Bergoglio, R.; Tello, E.E.; Nicolli, H.; and Smith, A.H. (1996.) Bladder cancer mortality associated with arsenic in drinking water in Argentina. *Epidemiology*, 7: 117-124.

Hsieh, L.L.; Chen, H.J.; Hsieh, J.T.; Jee, S.H.; Chen, G.S.; and Chen, C.J. (1994.) Arsenic-related Bowen's disease and paraquat-related skin cancerous lesions show no detectable ras and p53 gene alterations. *Cancer Lett.*, 86: 59-65.

Hsueh, Y.M.; Cheng, G.S.; Wu, M.M.; Yu, H.S.; Kuo, T.L.; and Chen, C.J. (1995.) Multiple risk factors associated with arsenic-induced skin cancer: Effects of chronic liver disease and malnutritional status. *Br. J. Cancer*, 71: 109-114.

Hu, J.L.; Guan, X.J.; and Sanchez, E.R. (1996.) Enhancement of glucocorticoid receptor-mediated gene expression by cellular stress: Evidence for the involvement of a heat shock-initiated factor or process during recovery from stress. *Cell Stress Chaperones*, 1: 197-205.

Hu, Y.; Su, L.; and Snow, E.T. (1998.) Arsenic toxicity is enzyme specific and its affects on ligation are not caused by the direct inhibition of DNA repair enzymes. *Mutat. Res.*, 408: 203-218.

Huang, R.N.; Ho, I.C.; Yih, L.H.; and Lee, T.C. (1995.) Sodium arsenite induces chromosome endoreduplication and inhibits protein phosphatase activity in human fibroblasts. *Environ. Mol. Mutagen*, 25: 188-196.

Huang, S.C., and Lee, T.C. (1998.) Arsenite inhibits mitotic division and perturbs spindle dynamics in HeLa S3 cells. *Carcinogenesis*, 19: 889-896.

Huot, J.; Lambert, H.; Lavoie, J.N.; Guimond, A.; Houle, F.; and Landry, J. (1995.) Characterization of 45-kDa/54-kDa HSP27 kinase, a stress-sensitive kinase which may activate the phosphorylation-dependent protective function of mammalian 27-kDa heat shock protein HSP27. *Eur. J. Biochem.*, 227: 416-427.

Ishikawa, T.; Igarashi, T.; Hata, K.; and Fujita, T. (1999.) *c-fos* induction by heat, arsenite, and cadmium is mediated by a heat shock element in its promoter. *Biochem. Biophys. Res. Commun.*, 254: 566-571.

Ito, H.; Okamoto, K.; Nakayama, H.; Isobe, T.; and Kato, K. (1997.) Phosphorylation of alphaB-crystallin in response to various types of stress. *J. Biol. Chem.*, 272: 29934-29941.

Jaafar, R.; Omar, I.; Jidon, A.J.; Wan-Khamizar, B.W.; Siti-Aishah, B.M.; and Sharifah-Noor-Akmal, S.H. (1993.) Skin cancer caused by chronic arsenical poisoning—A report of three cases. *Med. J. Malaysia*, 48: 86-92.

Jacob, A.K.; Hotchkiss, R.S.; DeMeester, S.L.; Hiramatsu, M.; Karl, I.E.; Swanson, P.E.; Cobb, J.P.; and Buchman, T.G. (1997.) Endothelial cell apoptosis is accelerated by inorganic iron and heat via an oxygen radical dependent mechanism. *Surgery*, 122: 243-253.

Jha, A.N.; Noditi, M.; Nilsson, R.; and Natarajan, A.T. (1992.) Genotoxic effects of sodium arsenite on human cells. *Mutat. Res.*, 284: 215-221.

Ju, W.D.; Schiller, J.T.; Kazempour, M.K.; and Lowy, D.R. (1993.) TGF alpha enhances locomotion of cultured human keratinocytes. *J. Invest. Dermatol.*, 100: 628-632.

Jurivich, D.A.; Pachetti, C.; Qiu, L.; and Welk, J.F. (1995.) Salicylate triggers heat shock factor differently than heat. *J. Biol. Chem.*, 270: 24489-24495.

Kampinga, H.H.; Brunsting, J.F.; Stege, G.J.; Burgman, P.W.; and Konings, A.W. (1995.) Thermal protein denaturation and protein aggregation in cells made thermotolerant by various chemicals: Role of heat shock proteins. *Exp. Cell. Res.*, 219: 536-546.

Karjalainen, S.; Salo, H.; and Teppo, L. (1989.) Basal cell and squamous cell carcinoma of the skin in Finland. Site distribution and patient survival. *Int. J. Dermatol.*, 28: 445-450.

Kato, K.; Ito, H.; and Okamoto, K. (1997.) Modulation of the arsenite-induced expression of stress proteins by reducing agents. *Cell Stress Chaperones*, 2: 199-209.

Kato, K.; Yamanaka, K.; Nakano, M.; Hasegawa, A.; Oku, N.; and Okada, S. (1997.) Cell-nuclear accumulation of 72-kDa stress protein induced by dimethylated arsenics. *Biol. Pharm. Bull.*, 20: 364-369.

Karagas, M.R.; Stukel, T.A.; Greenberg, E.R.; Baron, J.A.; Mott, L.A.; and Stern, R.S. (1992.) Risk of subsequent basal cell carcinoma and squamous cell carcinoma of the skin among patients with prior skin cancer. Skin Cancer Prevention Study Group. *JAMA*, 267(24): 3305-10.

Karagas, M.R.; Testeson, T.D.; Blum, J.D.; Klaue, B.; Weiss, J.E.; Stannard, V.; Spate, V.; and Morris, J.S. (2000.) Measurement of low levels of arsenic exposure: A comparison of water and toenail concentrations. *Am. J. Epid.*, 152: 84-90.

Karagas, M.R.; Testeson, T.D.; Blum, J.D.; Morris, J.S.; Baron, J.A.; and Klaue, B. (1998.) Design of an epidemiologic study of drinking water arsenic exposure and skin and bladder cancer risk in a U.S. population. *Environ. Health Perspect.*, 106 (Suppl. 4): 1047-1050.

Karnovsky, M.J. (1965.) A formaldehyde-gluteraldehyde fixative of high osmolarity for use in electron microscopy. *J. Cell Biol.*, 27: 137a.

Kawasaki, H.; Moriguchi, T.; Matsuda, S.; Li, H.Z.; Nakamura, S.; Shimohama, S.; Kimura, J.; Gotoh, Y.; and Nishida, E. (1996.) Ras-dependent and Ras-independent activation pathways for the stress-activated-protein-kinase cascade. *Eur. J. Biochem.*, 241: 315-321.

Kim, S.H.; Yeo, G.S.; Lim, Y.S.; Kang, C.D.; Kim, C.M.; and Chung, B.S. (1998.) Suppression of multidrug resistance via inhibition of heat shock factor by quercetin in MDR cells. *Exp. Mol. Med.*, 30: 87-92.

Kitchin, K.T. (2001.) Recent advances in arsenic carcinogenesis: Modes of action, animal model systems and methylated arsenic metabolites. *Toxicol. Appl. Pharmacol.*, 172: 249-261.

Kochhar, T.S.; Howard, W.; Hoffman, S.; and Brammer-Carleton, L. (1996.) Effect of trivalent and pentavalent arsenic in causing chromosome alterations in cultured Chinese hamster ovary (CHO) cells. *Toxicol. Lett.*, 84(1): 37-42.

Kodym, R.; Calkins, P.; and Story, M. (1999.) The cloning and characterization of a new stress response protein. A mammalian member of a family of theta class glutathione-S-transferase-like proteins. *J. Biol. Chem.*, 274: 5131-5137.

Konishi, H.; Matsuzaki, H.; Tanaka, M.; Takemura, Y.; Kuroda, S.; Ono, Y.; and Kikkawa, U. (1997.) Activation of protein kinase B (Akt/RAC-protein kinase) by cellular stress and its association with heat shock protein Hsp27. *FEBS Lett.*, 410: 493-498.

Korber, P.; Zander, T.; Herschlag, D.; and Bardwell, J.C.A. (1999.) A new heat shock protein that binds nucleic acids. *J. Biol. Chem.*, 274: 249-256.

Kovacs, A.; Yonemoto, K.; Katsuoka, K.; Nishiyama, S.; and Harhai, I. (1996.) Bowen's disease: Statistical study of a 10 year period. *J. Dermatol.*, 23: 267-74.

Ku, W.W., and Bernstein, I.A. (1988.) Lectin binding as a probe of proliferative and differentiative phases in primary monolayer cultures of cutaneous keratinocytes. *Exp. Cell Res.*, 175: 298-316.

Kültz, D., and Csonka, L. (1999.) What sets the TonE during osmotic stress? *Proc. Natl. Acad. Sci. USA*, 96: 1814-1816.

Lambert, H.; Charette, S.J.; Bernier, A.F.; Guimond, A.; and Landry, J. (1999.) HSP27 multimerization mediated by phosphorylation-sensitive intermolecular interactions at the amino terminus. *J. Biol. Chem.*, 274: 9378-9385.

Lappas, G.D.; Karl, I.E.; and Hotchkiss, R.S. (1994.) Effect of ethanol and sodium arsenite on HSP-72 formation and on survival in a murine endotoxin model. *Shock*, 2: 34-39.

Lee, T.C., and Ho, I.C. (1994.) Differential cytotoxic effects of arsenic on human and animal cells. *Environ. Health Perspect.*, 102 (Suppl. 3): 101-105.

Lee, T.C.; Tanaka, N.; Lamb, P.W.; Gilmer, T.M.; and Barrett, J.C. (1988.) Induction of gene amplification by arsenic. *Science*, 241: 79-81.

Lee, Y.J., and Corry, P.M. (1998.) Metabolic oxidative stress-induced HSP70 gene expression is mediated through SAPK pathway. Role of Bcl-2 and *c-jun* NH_2-terminal kinase. *J. Biol. Chem.*, 273: 29857-29863.

Leonard, A. (1991.) Metals and metalloids and their ions and compounds. In *Metals and Their Compounds in the Environment.* Edited by E. Merian. Basel: VCH. pp. 751-774.

Leong, J.; Hughes-Fulford, M.; Rakhlin, N.; Habib, A.; Maclouf, J.; and Goldyne, M.E. (1996.) Cyclooxygenases in human and mouse skin and cultured human keratinocytes: Association of COX-2 expression with human keratinocyte differentiation. *Exp. Cell Res.*, 224(1): 79-87.

Lerda, D. (1994.) Sister-chromatid exchange (SCE) among individuals chronically exposed to arsenic in drinking water. *Mutat. Res.*, 312: 111-120.

Levin, A.J. (1997.) p53, the cellular gatekeeper for growth and division. *Cell*, 88: 323-331.

Leyh, F., and Rothlaender, J.P. (1985.) Multiple bowenoide Arsen-Keratosen. *Derm. Beruf. Umwelt*, 33: 99-101.

Lim, C.P.; Jain, N.; and Cao, X. (1998.) Stress-induced immediate-early gene, *egr*-1, involves activation of p38/JNK1. *Oncogene*, 16: 2915-2926.

Liu, R.Y.; Corry, P.M.; and Lee, Y.J. (1995.) Potential involvement of a constitutive heat shock element binding factor in the regulation of chemical stress-induced hsp70 gene expression. *Mol. Cell Biochem.*, 144: 27-34.

Liu, S.X.; Athar, M.; Lippai, I.; Waldren, C.; and Hei, T.K. (2001.) Induction of oxyradicals by arsenic: Implication for mechanism of genotoxicity. Published online before print February 6, 2001. *Proc. Natl. Acad. Sci. USA*, 101: 1073-1084.

Lodhi, I.J.; Sweeney, J.F.; Clift, R.E.; and Hinshaw, D.B. (2001.) Nuclear dependence of sulfur mustard-mediated cell death. *Toxicol. Appl. Pharmacol.*, 170: 69-77.

Lowe, S.W.; Schmitt, E.M.; Smith, B.A.; Osborne, B.A.; and Jacks, T. (1993.) P53 is required for radiation induced apoptosis in mouse thymocytes. *Nature*, 362: 847-852.

Lowry, O.H.; Rosenbrough, N.H.; Farr, A.L.; and Randall, R.J. (1951.) Protein measurement with the folin phenol reagent. *J. Biol. Chem.*, 193: 265-275.

Luchtrath, H. (1983.) The consequences of chronic arsenic poisoning among Moselle wine growers: Pathoanatomical investigations of post-mortem examination performed between 1960 and 1972. *J. Cancer Res. Clin. Oncol.*, 105, 172-182.

Ludwig, S.; Hoffmeyer, A.; Goebeler, M.; Kilian, K.; Hafner, H.; Neufeld, B.; Han, J.; and Rapp, U.R. (1998.) The stress inducer arsenite activates mitogen-activated protein kinases extracellular signal-regulated kinases 1 and 2 via a MAPK kinase 6/p38-dependent pathway. *J. Biol. Chem.*, 23;273(4): 1917-1922.

Lynn, S.; Shiung, J.N.; Gurr, J.R.; and Jan, K.Y. (1998.) Arsenite stimulates poly(ADP-ribosylation) by generation of nitric oxide. *Free Radical Biol. Med.*, 24: 442-449.

Macfarlane, W.M.; Smith, S.B.; James, R.F.; Clifton, A.D.; Doza, Y.N.; Cohen, P.; and Docherty, K. (1997.) The p38/reactivating kinase mitogen-activated protein kinase cascade mediates the activation of the transcription factor insulin upstream factor 1 and insulin gene transcription by high glucose in pancreatic beta-cells. *J. Biol. Chem.*, 272: 20936-20944.

Mandal, B.K.; Agra, Y.; and Suzuki, K.T. (2001.) Identification of dimethylarsinous and monomethylarsonous acids in human urine of the arsenic-affected areas of West Bengal, India. *Chem. Res. Toxicol.*, 14: 371-378.

Mass, M.J., and Wang, L. (1997.) Arsenic alters cytosine methylation patterns of the promoter of the tumor suppressor gene p53 in human lung cells: A model for a mechanism of carcinogenesis. *Mutat. Res.*, 386(3): 263-77.

Masuya, Y.; Hioki, K.; Tokunaga, R.; and Taketani, S. (1998.) Involvement of the tyrosine phosphorylation pathway in induction of human heme oxygenase-1 by hemin, sodium arsenite, and cadmium chloride. *J. Biochem. (Tokyo)*, 124: 628-633.

Maytin, E.V. (1992.) Differential effects of heat shock and UVB light upon stress protein expression in epidermal keratinocytes. *J. Biol. Chem.*, 267: 23189-23196.

Mazereeuw-Hautier, J.; Redoules, D.; Tarroux, R.; Charveron, M.; Salles, J.P.; Simon, M.F.; Cerutti, I.; Assalit, M.F.; Gall, Y.; Bonafe, J.L.; and Chap, H. (2000.) Identification of pancreatic type I secreted phospholipase A2 in human epidermis and its determination by tape stripping. *Br. J. Dermatol.*, 142(3): 424-31.

McCabe, M.; Maguire, D.; and Nowak, M. (1983.) The effects of arsenic compounds on human and bovine lymphocyte mitogenesis in vitro. *Environ. Res.*, 31: 323-331.

McDowell, H.E.; Eyers, P.A.; and Hundal, H.S. (1998.) Regulation of system A amino acid transport in L6 rat skeletal muscle cells by insulin, chemical and hyperthermic stress. *FEBS Lett.*, 441: 15-19.

Meng, Z., and Hsie, A.W. (1996.) Polymerase chain reaction-based deletion analysis of spontaneous and arsenite-enhanced gpt mutants in CHO-AS52 cells. *Mutat. Res.*, 356(2): 255-9.

Meng, Z., and Meng, N. (1994.) Effects of inorganic arsenicals on DNA synthesis in unsensitized human blood lymphocytes in vitro. *Biol. Trace Elem. Res.*, 42(3): 201-208.

Mirkes, P.E.; Doggett, B.; and Cornel, L. (1994.) Induction of a heat shock response (HSP 72) in rat embryos exposed to selected chemical teratogens. *Teratology*, 49: 135-142.

Mivechi, N.F.; Koong, A.C.; Giaccia, A.J.; and Hahn, G.M. (1994.) Analysis of HSF-1 phosphorylation in A549 cells treated with a variety of stresses. *Int. J. Hyperthermia*, 10: 371-379.

Miyakawa, H.; Woo, S.K.; Dahl, S.C.; Handler, J.S.; and Kwon, H.M. (1999.) Tonicity-responsive enhancer binding protein, a Rel-like protein that stimulates transcription in response to hypertonicity. *Proc. Natl. Acad. Sci. USA*, 96: 2538-2542.

Moore, L.E.; Huang, W.I.; Chung, J.; and Hayes, R.B. (2003.) Epidemiologic considerations to assess DNA methylation from environmental exposures in cancer. *Annals NY Acad. Sci.*, 983: 181-186.

Moore, L.E.; Smith, A.H.; Hopenhayn-Rich, C.; Biggs, M.L.; Kalman, D.A.; and Smith, M.T. (1997.) Micronuclei in exfoliated bladder cells among individuals chronically exposed to arsenic in drinking water. *Cancer Epidemiol. Biomarkers Prev.*, 6: 31-36.

Mosser, D.D.; Duchaine, J.; and Massie, B. (1993.) The DNA-binding activity of the human heat shock transcription factor is regulated in vivo by hsp70. *Mol. Cell. Biol.*, 13: 5427-5438.

Muller-Decker, K.; Heinzelmann, T.; Furstenberger, G.; Kecskes, A.; Lehmann, W.D.; and Marks, F. (1998.) Arachidonic acid metabolism in primary irritant dermatitis produced by patch testing of human skin with surfactants. *Toxicol. Appl. Pharmacol.*, 153(1): 59-67.

Muramatsu, T.; Kobayashi, N.; Tada, H.; Hatoko, M.; and Shirai, T. (1995.) Induction of the 72-kD heat shock protein in human skin melanoma and squamous cell carcinoma cell lines. *J. Dermatol.*, 22: 907-912.

Muramatsu, T.; Tada, H.; Kobayashi, N.; Yamji, M.; Shirai, T.; and Ohnishi, T. (1992.) Induction of the 72-kD heat shock protein in organ-cultured normal human skin. *J. Invest. Dermatol.*, 98: 786-790.

Nakamuro, K., and Sayato, Y. (1981.) Comparative studies of chromosomal aberration induced by trivalent and pentavalent arsenic. *Mutat. Res.*, 88: 73-80.

NAS. (1999.) *Arsenic in Drinking Water.* Washington, D.C.: National Academy of Sciences Press.

Newman, D.K.; Kennedy, E.K.; Coates, J.D.; Ahmann, D.; Ellis, D.J.; Lovley, D.R.; and Morel, F.M. (1997.) Dissimilatory arsenate and sulfate reduction in *Desulfotomaculum auripigmentum* sp. nov. *Arch. Microbiol.*, 168: 380-388.

Nriagu, J.O., ed. (1994.) *Arsenic in the Environment.* New York: John Wiley & Sons.

Nriagu, J.O.; Lawson, G.; Wong, H.K.T.; and Azcue, J.M. (1993.) A protocol for minimizing contamination in the analysis of trace metals in the Great Lakes. *J. Great Lakes Res.*, 19: 175-182.

Ochi, T.; Nakajima, F.; and Fukumori, N. (1998.) Different effects of inorganic and dimethylated arsenic compounds on cell morphology, cytoskeletal organization, and DNA synthesis in cultured Chinese hamster V79 cells. *Arch. Toxicol.*, 72: 566-573.

Ohan, N.W.; Tam, Y.; Fernando, P.; and Heikkila, J.J. (1998.) Characterization of a novel group of basic small heat shock proteins in *Xenopus laevis* A6 kidney epithelial cells. *Biochem. Cell Biol.*, 76(4): 665-671.

Ohnishi, Y.; Murakami, S.; Ohtsuka, H.; Miyauchi, S.; Shinmori, H.; and Hashimoto, K. (1997.) Merkel cell carcinoma and multiple Bowen's disease: Incidental association or possible relationship to inorganic arsenic exposure? *J. Dermatol.*, 24: 310-316.

Olive, P.L.; Horsman, M.R.; Grau, C.; and Overgaard, J. (1997.) Detection of hypoxic cells in a C3H mouse mammary carcinoma using the comet assay. *Br. J. Cancer*, 76(6): 694-649.

Ostling, O., and Johanson, K.J. (1984.) Microelectrophoretic study of radiation-induced DNA damages in individual mammalian cells. *Biochem. Biophys. Res. Commun.*, 123: 291-298.

Ovelgonne, H.; Bitorina, M.; and Van Wijk, R. (1995.) Stressor-specific activation of heat shock genes in H35 rat hepatoma cells. *Toxicol. Appl. Pharmacol.*, 135: 100-109.

Parker, M., and Murphy, M. (1995.) Peroxynitrite formed by simultaneous nitric oxide and superoxide generation causes cyclosporin-a sensitive mitochondrial efflux and depolarization. *Eur. J. Biochem.*, 234: 231-239.

Plappert, U.G.; Stocker, B.; Fender, H.; and Fliedner, T.M. (1997.) Changes in the repair capacity of blood cells as a biomarker for chronic low-dose exposure to ionizing radiation. *Environ. Mol. Mutagen.*, 30: 153-160.

Pontius, F.W.; Brown, K.G.; and Chen, C.J. (1994); Health implications of arsenic in drinking water. *Jour. AWWA*, 86: 52-63.

Pu, Y.P, and Bernstein, I.A. (1995.) Use of human pseudoepidermis to evaluate the toxicity of bis-(2-chloroethyl)sulfide (BCES) on water permeation barrier formation and function. *Toxicology Lett.*, 76: 85-91.

Pu, Y.P.; Bernstein, I.A.; Bernstam, L.; and Bronaugh, R.L. (1995.) Growing a stratified cornified primary culture of rat keratinocytes with epidermis-like water permeation barrier function. *In Vitro Cell Dev. Biol.*, 31: 283-287.

Radha, S., and Natarajan, A.T. (1998.) Sodium arsenite-induced chromosomal aberrations in the Xq arm of Chinese hamster cell lines. *Mutagenesis*, 13: 229-234.

Rahman, M.S.; Hull, L.L.; and Hughes, M.F. (1994.) In vitro percutaneous absorption of sodium arsenate in B6C3F1 mice. *Toxicology in vitro*, 8: 441-448.

Ramirez, P.; Eastmond, D.A.; Laclette, J.P.; and Ostrosky-Wegman, P. (1997.) Disruption of microtubule assembly and spindle formation as a mechanism for the induction of aneuploid cells by sodium arsenite and vanadium pentoxide. *Mutat. Res.*, 386: 291-298.

Ribeiro, P.L.; Mitra, R.S.; and Bernstein, I.A. (1991.) Assessment of the role of DNA damage and repair in the survival of primary cultures of rat cutaneous keratinocytes exposed to bis(2-chloroethyl)sulfide. *Toxicol. Appl. Pharmacol.*, 111: 342-351.

Ribeiro, S.P.; Villar, J.; Downey, G.P.; Edelson, J.D.; and Slutsky, A.S. (1994.) Sodium arsenite induces heat shock protein-72 kilodalton expression in the lungs and protects rats against sepsis. *Crit. Care Med.*, 22: 922-999.

Roat, J.W.; Wald, A.; Mendelow, H.; and Pataki, K.I. (1982.) Hepatic angiosarcoma associated with short-term arsenic ingestion. *Am. J. Med.*, 73: 933-936.

Rossman, T.; Meyn, M.S.; and Troll, W. (1975.) Effects of sodium arsenite on the survival of UV-irradiated *Escherichia coli*: Inhibition of a recA-dependent function. *Mutat. Res.*, 30(2): 157-62.

Rossman, T.G.; Goncharova, E.I.; Rajah, T.; and Wang, Z. (1997.) Human cells lack the inducible tolerance to arsenite seen in hamster cells. *Mutat. Res.*, 386: 307-314.

Rouse, J.; Cohen, P.; Trigon, S.; Morange, M.; Alonso-Llamazares, A.; Zamanillo, D.; Hunt, T.; and Nebreda, A.R. (1994.) A novel kinase cascade triggered by stress and heat shock that stimulates MAPKAP kinase-2 and phosphorylation of the small heat shock proteins. *Cell*, 78: 1027-1037.

Rudel, R.; Slayton, T.M.; and Beck, B.D. (1996.) Implications of arsenic genotoxicity for dose response of carcinogenic effects. *Regul. Toxicol. Pharmacol.*, 23(2): 87-105.

Rupa, D.S.; Schuler, M.; and Eastmond, D.A. (1997.) Detection of hyperdiploidy and breakage affecting the 1cen-1q12 region of cultured interphase human lymphocytes treated with various genotoxic agents. *Environ. Mol. Mutagen.*, 29: 161-167.

Rys-Sikora, K.E.; Konger, R.L.; Schoggins, J.W.; Malaviya, R.; and Pentland, A.P. (2000.) Prostaglandin E(2) regulation of cyclooxygenase expression in keratinocytes is mediated via cyclic nucleotide-linked prostaglandin receptors. *J. Lipid Res.*, 4: 873-881.

Sacco, M.G.; Zecca, L.; Bagnasco, L.; Chiesa, G.; Parolini, C.; Bromley, P.; Cato, E.M.; Roncucci, R.; Clerici, L.A.; and Vezzoni, P. (1997.) A transgenic mouse model for the detection of cellular stress induced by toxic inorganic compounds. *Nat. Biotechnol.*, 15: 1392-1397.

Salazar, A.M.; Ostrosky-Wegman, P.; Menendez, D.; Miranda, E.; Garcia-Carranca, A.; and Rojas, E. (1997.) Induction of p53 protein expression by sodium arsenite. *Mutat. Res.*, 381(2): 259-265.

Samet, J.; Graves, L.; Quay, J.; Dailey, L.; Devlin, R.; Ghio, A.; Wu, W.; Bromberg, P.; and Reed, W. (1998.) Activation of MAPKs in human bronchial epithelial cells exposed to metals. *AJP*, 275: 3.

Sanchez, E.R.; Hu, J.L.; Zhong, S.; Shen, P.; Greene, M.J.; and Housley, P.R. (1994.) Potentiation of glucocorticoid receptor-mediated gene expression by heat and chemical shock. *Mol. Endocrinol.*, 8: 408-421.

Scavarelli-Karantsavelos, R.M.; Zaman-Saroya, S.; Vaughan, F.L.; and Bernstein, I.A. (1990.) A pseudo-epidermis constructed *in vitro*, for use in toxicological and pharmacological studies. *J. Skin Physiol.*, 3: 115-125.

Schadow, A.; Scholz-Pedretti, K.; Lambeau, G.; Gelb, M.H.; Furstenberger, G.; Pfeilschifter, J.; and Kaszkin, M. (2001).Characterization of group X phospholipase A(2) as the major enzyme secreted by human keratinocytes and its regulation by the phorbol ester TPA. *J. Invest. Dermatol.*, 116: 31-39.

Schett, G.; Redlich, K.; Xu, Q.; Bizan, P.; Grger, M.; Tohidast-Akrad, M.; Kiener, H.; Smolen, J.; and Steiner, G. (1998.) Enhanced expression of heat shock protein 70 (hsp70) and heat shock factor 1 (HSF1) activation in rheumatoid arthritis synovial tissue differential regulation of hsp70 expression and HSF1 activation in synovial fibroblasts by proinflammatory cytokines, shear stress, and antiinflammatory drugs. *J. Clin. Invest.*, 102: 302-311.

Schwerdtle, T.; Walter, I.; Mackiw, I.; and Hartwig, A. (2003.) Induction of oxidative DNA damage by arsenite and its trivalent and pentavalent methylated metabolites in cultured human cells and isolated DNA. *Carcinogenesis*, 24: 967-974.

Senisterra, G.A.; Huntley, S.A.; Escaravage, M.; Sekhar, K.R.; Freeman, M.L.; Borrelli, M.; and Lepock, J.R. (1997.) Destabilization of the Ca2+-ATPase of sarcoplasmic reticulum by thiol-specific, heat shock inducers results in thermal denaturation at 37 degrees C. *Biochemistry*, 36: 11002-11011.

Setaro, F., and Morley, C.G.D. (1976.) A modified fluorimetric method for the determination of microgram quantities of DNA from cell or tissue cultures. *Anal. Biochem.*, 71: 313-317.

Shannon, R.L., and Strayer, D.S. (1989.) Arsenic-induced skin toxicity. *Human Toxicol.*, 8: 99-104.

Shaw, M.; Cohen, P.; and Alessi, D.R. (1998.) The activation of protein kinase B or heat shock is mediated by phosphoinositide 3-kinase and not by mitogen-activated protein kinase-activated protein kinase-2. *Biochem. J.*, 336: 241-246.

Shay, J.W. (1995.) Aging and cancer: Are telomeres and telomerase the connection? *Mol. Med. Today*, Nov. 1(8): 378-384.

Shen, P.; Xie, Z.J.; Li, H.; and Sanchez, E.R. (1993.) Glucocorticoid receptor conversion to high affinity nuclear binding and transcription enhancement activity in Chinese hamster ovary cells subjected to heat and chemical stress. *J. Steroid Biochem. Mol. Biol.*, 47: 55-64.

Shi, B., and Isseroff, R.R. (1996.) Thapsigargin induces phosphorylation of the 27-kDa heat shock protein in human keratinocytes. *J. Invest. Dermatol.*, 107: 749-54.

Shibata, M.A.; Ward, J.M.; Green, J.E.; and Merlino, G. (1997.) Enhanced sensitivity to tumor growth and development in multistage skin carcinogenesis by transforming growth factor-alpha-induced epidermal growth factor receptor activation but not p53 inactivation. *Mol. Carcinog.*, 18(3): 160-70.

Shimizu, M.; Hochadel, J.F.; Fulmer, B.A.; and Waalkes, M.P. (1998.) Effect of glutathione depletion and metallothionein gene expression on arsenic-induced cytotoxicity and c-myc expression in vitro. *Toxicol. Sci.*, 45: 204-11.

Siebenlist, U.; Franzoso, G.; and Brown, K. (1994.) Structure, regulation and function of NF-kappa B. *Annu. Rev. Cell Biol.*, 10: 405-55.

Silver, S. (1998.) Genes for all metals: A bacterial view of the periodic table. The 1996 Thom Award Lecture. *J. Ind. Microbiol. Biotechnol.*, 20: 1-12.

Simeonova, P.; Wang, S.; Hulderman, T.; and Luster, M. (2002.) C-Src-dependent activation of the epidermal growth factor receptor and mitogen-activated protein kinase pathway by arsenic. *J. Biol. Chem.*, 27: 4.

Simeonova, P.; Wang, S.; Kashon, M.; Kommineni, C.; Crecelius, E.; and Luster, M. (2001.) Quantitative relationship between arsenic exposure and AP-1 activity in mouse urinary bladder epithelium. *Toxicol. Sci.*, 60: 2.

Simeonova, P.; Wang, S.; Toriuma, W.; Kommineni, V.; Matheson, J.; Unimye, N.; Kayama, F.; Harki, D.; Ding, M.; Vallyathan, V.; and Luster, M. (2000.) Arsenic mediates cell proliferation and gene expression in the bladder epithelium: Association with activating protein-1 transactivation. *Cancer Res.*, 60: 13.

Singh, N.P.; McCoy, M.T.; Tice, R.R.; and Schneider, E.L. (1988.) A simple technique for quantitation of low levels of DNA damage in individual cells. *Exp. Cell Res.*, 175: 184-191.

Sjursen, W.; Brekke, O.L.; and Johansen, B. (2000.) Secretory and cytosolic phospholipase A(2)regulate the long-term cytokine-induced eicosanoid production in human keratinocytes. *Cytokine*, 12: 1189-94.

Smith, A.H.; Arroyo, A.P.; Mazumder, D.N.; Kosnett, M.J.; Hernandez, A.L.; Beeris, M.; Smith, M.T.; and Moore, L.E. (2000.) Arsenic induced skin lesions among Atacameno people in Northern Chile despite good nutrition and centuries of exposure. *Environ. Health Prospect.*, 108: 617-620.

Smith, A.H.; Hopenhayn-Rich, C.; Bates, M.N.; Goeden, H.M.; Hertz-Picciotto, I.; Duggan, H.M.; Wood, R.; Kosnett, M.J.; and Smith, M.T. (1992.) Cancer risks from arsenic in drinking water. *Environ. Health Perspect.*, 97: 259-267.

Suzuki, A.; Kozawa, O.; Oiso, Y.; and Kato, K. (1996.) Protein kinase C activation inhibits stress-induced synthesis of heat shock protein 27 in osteoblast-like cells: Function of arachidonic acid. *J. Cell. Biochem.*, 62: 69-75.

Suzuki, K.; Wakao, N.; Sakurai, Y.; Kimura, T.; Sakka, K.; and Ohmiya, K. (1997.) Transformation of *Escherichia coli* with a large plasmid of *Acidiphilium multivorum* AIU 301 encoding arsenic resistance. *Appl. Environ. Microbiol.*, 63: 2089-2091.

Takechi, H.; Hosokawa, N.; Hirayoshi, K.; and Nagata, K. (1994.) Alternative 5' splice site selection induced by heat shock. *Mol. Cell Biol.*, 4(1): 567-575.

Tamura, Y.; Tsuboi, N.; Sato, N.; and Kikuchi, K. (1993.) 70 kDa heat shock cognate protein is a transformation-associated antigen and a possible target for the hosts anti-tumor immunity. *J. Immunol.*, 151: 5516-5524.

Tanabe, M.; Nakai, A.; Kawazoe, Y.; and Nagata, K. (1997.) Different thresholds in the responses of two heat shock transcription factors, HSF1 and HSF3. *J. Biol. Chem.*, 272: 15389-15395.

Tates, A.D.; van Dam, F.J.; de Zwart, F.A.; Darroudi, F.; Natarajan, A.T.; Rossner, P.; Taylor, R.S.; Ramirez, R.D.; Ogoshi, M.; Chaffins, M.; Piatyszek, M.A.; and Shay, J.W. (1996.) Detection of telomerase activity in malignant and nonmalignant skin conditions. *J. Invest. Dermatol.*, 106: 759-765.

Tay, C.H., and Seah, C.S. (1975.) Arsenic poisoning from anti-asthmatic herbal preparations. *Med. J. Aust.*, 2: 424-428.

Taylor, R.S.; Ramirez, R.D.; Ogoshi, M.; Chaffins, M.; Piatyszek, M.A.; and Shay, J.W. (1996.) Detection of telomerase activity in malignant and nonmalignant skin conditions. *J. Invest. Dermatol.*, 106: 759-65.

Tello, E.E. (1988.) Carcinomas of internal organs and their relationship to arsenical drinking water in the Republic of Argentina. *Med. Cutan. Ibero Lat. Am.*, 16: 497-501.

Thomas, D.J.; Styblo, M.; and Lin, S. (2001). The cellular metabolism and systemic toxicity of arsenic. *Toxicol. Appl. Pharmacol.*, 176(2): 127-44.

Thornton, I. (1996.) Sources and pathways of arsenic in the geochemical environment: Health implications. In *Environmental Geochemistry and Health.* Edited by J.D. Appleton, R. Fuge, and G.J.H. McCall. London: Geological Society Special Publications No. 113. pp. 153-161.

Tice, R.R.; Yager, J.W.; Andrews, P.; and Crecelius, E. (1997.) Effect of hepatic methyl donor status on urinary excretion and DNA damage in B6C3F1 mice treated with sodium arsenite. *Mutat. Res.*, 386(3): 315-34.

Tsuda, T.; Babazono, A.; Yamamoto, E.; Kurumatani, N.; Mino, Y.; Ogawa, T.; Kishi, Y.; and Aoyama, H. (1995.) Ingested arsenic and internal cancer: A historical cohort study followed for 33 years. *Am. J. Epidemiol.*, 141: 198-209.

Upshaw Jr., C.B.; Bryant, M.F.; and Claiborne Jr., T.S. (1979.) Noncirrhotic portal hypertension after arsenic ingestion. *South Med. J.*, 72: 1332-1334.

USEPA (U.S. Environmental Protection Agency). (1992.) *Dermal Exposure Assessment: Principles and Applications.* Report #EPA/600/8-91/011B. Washington, D.C.: USEPA, Office of Health and Environmental Assessment.

USEPA (U.S. Environmental Protection Agency). (1996.) *Exposure Factors Handbook.* Report #EPA/600/P-95/002Ba. Washington, D.C.: USEPA, Office of Research & Development.

USEPA (U.S. Environmental Protection Agency). (2001.) *Fed. Reg.*, Jan. 22, 2001.

Vahter, M. (1999.) Methylation of inorganic arsenic in different mammalian species and population groups. *Sci. Prog.*, 82: 69-88.

van Delft, S.; Coffer, P.; Kruijer, W.; and van Wijk, R. (1993.) *c-fos* induction by stress can be mediated by the SRE. *Biochem. Biophys. Res. Commun.*, 197: 542-548.

Van Goethem, F.; Lison, D.; and Kirsch-Volders, M. (1997.) Comparative evaluation of the in vitro micronucleus test and the alkaline single cell gel electrophoresis assay for the detection of DNA damaging agents: Genotoxic effects of cobalt powder, tungsten carbide and cobalt-tungsten carbide. *Mutat. Res.*, 392: 31-43.

Vaughan, F.L.; Gray, R.H.; and Bernstein, I.A. (1986.) Growth and differentiation of primary rat keratinocytes on synthetic membranes. *In Vitro Devel. Cell Biol.*, 22: 141-149.

Vega, L.; Gonsebatt, M.E.; and Ostrosky-Wegman, P. (1995.) Aneugenic effect of sodium arsenite on human lymphocytes in vitro: An individual susceptibility effect detected. *Mutat. Res.*, 334: 365-73.

Vietor, I., and Vilcek, J. (1994.) Pathways of heat shock protein 28 phosphorylation by TNF in human fibroblasts. *Lymphokine Cytokine Res.*, 13: 315-323.

Vig, B.K.; Figueroa, M.L.; Cornforth, M.N.; and Jenkins, S.H. (1984.) Chromosome studies in human subjects chronically exposed to arsenic in drinking water. *Am. J. Int. Med.*, 6: 325-338.

Villani, P.; Andreoli, C.; Crebelli, R.; Pacchierotti, F.; Zijno, A.; and Carere, A. (1998.) Analysis of micronuclei and DNA single-strand breaks in mouse splenocytes and peripheral lymphocytes after oral administration of tetramethylthiuram disulfide (thiram). *Food Chem. Toxicol.*, 36: 155-164.

Wagner, S.L.; Maliner, J.S.; Morton, W.E.; and Braman, R.S. (1979.) Skin cancer and arsenical intoxication from well water. *Arch. Dermatol.*, 115: 1205-1207.

Wang, B.J.; Lee, Y.Y.; Mak, C.P.; Kao, H.F.; Hsu, M.L.; and Hsien, J.R. (1991.) Quantitative and morphological changes of Langerhans cells in Bowen's disease from patients with chronic arsenicism. *J. Formos Med. Assoc.*, 90: 1093-1098.

Wang, J.H.; Redmond, H.P.; Watson, R.W.; and Bouchier-Hayes, D. (1997.) Induction of human endothelial cell apoptosis requires both heat shock and oxidative stress responses. *Am. J. Physiol.*, 272(5 Pt 1): C1543-1551.
Wang, J.H.; Redmond, H.P.; Watson, R.W.; Condron, C.; and Bouchier-Hayes, D. (1995.) Induction of heat shock protein 72 prevents neutrophil-mediated human endothelial cell necrosis. *Arch. Surg.*, 130: 1260-1265.
Wang, Z.; Dey, S.; Rosen, B.P.; and Rossman, T.G. (1996.) Efflux-mediated resistance to arsenicals in arsenic-resistant and hypersensitive Chinese hamster cells. *Toxicol. Applied Pharmacol.*, 137: 112-119.
Wang, Z.; Hou, G.; and Rossman, T.G. (1994.) Induction of arsenite tolerance and thermotolerance by arsenite occur by different mechanisms. *Environ. Health Perspect.*, 102 (Suppl. 3): 97-100.
Wang, Z.G.; Rivi, R.; Delva, L.; Konig, A.; Scheinberg, D.A.; Gambacorti-Passerini, C.; Gabrilove, J.L.; Warrell Jr., R.P.; and Pandolfi, P.P. (1998.) Arsenic trioxide and melarsoprol induce programmed cell death in myeloid leukemia cell lines and function in a PML and PML-RARalpha independent manner. *Blood*, 92: 1497-1504.
Watson, R.W.; Redmond, H.P.; Wang, J.H.; and Bouchier-Hayes, D. (1996.) Mechanisms involved in sodium arsenite-induced apoptosis of human neutrophils. *J. Leukoc. Biol.*, 60: 625-632.
Wester, R.C.; Maibach, H.I.; Sedik, L.; Melendres, J.; and Wade, M. (1993.) In vivo and in vitro percutaneous absorption and skin decontamination of arsenic in water and soil. *Fundam. Appl. Toxicol.*, 20: 336-340.
WHO (World Health Organization). (1993.) *Guidelines for Drinking Water Quality.* Geneva: WHO.
Wiegant, F.A.; Souren, J.E.; van Rijn, J.; and van Wijk, R. (1994.) Stressor-specific induction of heat shock proteins in rat hepatoma cells. *Toxicology*, 94: 143-159.
Wiegant, F.A.; Spieker, N.; Van der Mast, C.A.; and Van Wijk, R. (1996.) Is heat shock protein re-induction during tolerance related to the stressor-specific induction of heat shock proteins? *J. Cell. Physiol.*, 169: 364-372.
Wiegant, F.A.; Spieker, N.; and van Wijk, R. (1998.) Stressor-specific enhancement of hsp induction by low doses of stressors in conditions of self- and cross-sensitization. *Toxicology*, 127: 107-119.
Wijeweera, J.B.; Thomas, C.M.; Gandolfi, A.J.; and Brendel, K. (1995.) Sodium arsenite and heat shock induce stress proteins in precision-cut rat liver slices. *Toxicology*, 104: 35-45.
Winski, S.L., and Carter, D.E. (1998.) Arsenic toxicity in human keratinocytes: Characterization of morphologic changes and determination of the mechanism of damage. *J. Toxicol. Environ. Health Series A*, 53: 345-355.
Wong, H.R.; Finder, J.D.; Wasserloos, K.; and Pitt, B.R. (1995.) Expression of iNOS in cultured rat pulmonary artery smooth muscle cells is inhibited by the heat shock response. *Am. J. Physiol.*, 269: L843-L848.
Wong, H.R.; Mannix, R.J.; Rusnak, J.M.; Boota, A.; Zar, H.; Watkins, S.C.; Lazo, J.S.; and Pitt, B.R. (1996.) The heat-shock response attenuates lipopolysaccharide-mediated apoptosis in cultured sheep pulmonary artery endothelial cells. *Am. J. Respir. Cell Mol. Biol.*, 15: 745-751.
Wong, S.S.; Tan, K.C.; and Goh, C.L. (1998.) Cutaneous manifestations of chronic arsenicism: Review of seventeen cases. *J. Am. Acad. Dermatol.*, 38: 179-185.
Xu, C.; Zhou, T.; Kuroda, M.; and Rosen, B.P. (1998.) Metalloid resistance mechanisms in prokaryotes. *J. Biochem. (Tokyo)*, 123: 16-23.

Yager, J.W., and Wiencke, J.K. (1997.) Inhibition of poly(ADP-ribose) polymerase by arsenite. *Mutat. Res.*, 386: 345-351.

Yang, C., and Frenkel, K. (2002.) Arsenic-mediated cellular signal transduction, transcription factor actovation, and aberrant gene expression: Implications in carcinogenesis. *J. Environ. Pathol. Toxicol. Oncol.*, 21: 4.

Yang, Y.L.; Lu, K.T.; Tsay, H.J.; Lin, C.H.; and Lin, M.T. (1998.) Heat shock protein expression protects against death following exposure to heatstroke in rats. *Neurosci. Lett.*, 252: 9-12.

Yasuda, K.; Nakai, A.; Hatayama, T.; and Nagata, K. (1995.) Cloning and expression of murine high molecular mass heat shock proteins, HSP105. *J. Biol. Chem.*, 270: 29718-29723.

Yih, L.H.; Ho, I.C.; and Lee, T.C. (1997.) Sodium arsenite disturbs mitosis and induces chromosome loss in human fibroblasts. *Cancer Res.*, 57: 5051-5059.

Yih, L.H., and Lee, T.C. (1999.) Effects of exposure protocols on induction of kinetochore-plus and -minus micronuclei by arsenite in diploid human fibroblasts. *Mutat. Res.*, 440: 75-82.

Yu, H.S.; Chang, K.L.; Wang, C.M.; and Yu, C.L. (1992.) Alterations of mitogenic responses of mononuclear cells by arsenic in arsenical skin cancers. *J. Dermatol.*, 19: 710-714.

Yu, H.S.; Chiou, K.S.; Chen, G.S.; Yang, R.C.; and Chang, S.F. (1993.) Progressive alterations of cytokeratin expressions in the process of chronic arsenism. *J. Dermatol.*, 20: 741-745.

Yu, R.C.; Hsu, K.H.; Chen, C.J.; and Froines, J.R. (2000.) Arsenic methylation capacity and skin cancer. *Cancer Epid. Biomarkers Prevent.*, 9: 1259-1262.

Zaldivar, R.; Prunes, L.; and Ghai, G.L. (1981.) Arsenic dose in patients with cutaneous carcinomata and hepatic hemangio-endothelioma after environmental and occupational exposure. *Arch. Toxicol.*, 47: 145-154.

Zavialov, A.; Benndorf, R.; Ehrnsperger, M.; Zavyalov, V.; Dudich, I.; Buchner, J.; and Gaestel, M. (1998.) The effect of the intersubunit disulfide bond on the structural and functional properties of the small heat shock protein Hsp25. *Int. J. Biol. Macromol.*, 22: 163-173.

Zhao, C.Q.; Young, M.R.; Diwan, B.A.; Coogan, T.P.; and Waalkes, M.P. (1997.) Association of arsenic-induced malignant transformation with DNA hypomethylation and aberrant gene expression. *Proc. Natl. Acad. Sci. USA*, 94(20): 10907-10912.

Zhitkovich, A.; Shrager, S.; and Messer, J. (2000.) Reductive metabolism of Cr(VI) by cysteine leads to the formation of binary and ternary Cr—DNA adducts in the absence of oxidative DNA damage. *Chem. Res. Toxicol. 2000*, 13(11): 1114-1124.

Zhong, M.; Orosz, A.; and Wu, C. (1998.) Direct sensing of heat and oxidation by Drosophila heat shock transcription factor. *Mol. Cell*, 2: 101-108.

Zou, J.; Salminen, W.F.; Roberts, S.M.; and Voellmy, R. (1998.) Correlation between glutathione oxidation and trimerization of heat shock factor 1, an early step in stress induction of the Hsp response. *Cell Stress Chaperones*, 3: 130-141.

ABBREVIATIONS

BPE	bovine pituitary extract
DAB	diaminobenzidine tetrahydrochloride
DNA	deoxyribonucleic acid
EBSS	Earle's balanced salt solution
EGF	epidermal growth factor
ELISA	enzyme-linked immunosorbent assay
FBS	fetal bovine serum
GFAAS	graphite furnace atomic absorption spectrometry
HF	human fibroblasts
HK	human keratinocytes
K_p	permeability constant for arsenic
$K_{p,w}$	permeation constant for water
LDH	lactate dehydrogenase
MEM	minimum essential medium
MTS	tetrazolium compound [3-(4,5-dimethylthiozol-2-yl)-5-(3-carboxymethoxyphenyl)-2-(4-sulfophenyl)-2H-tetrazolium]
NAD	nicotin amide adenine dinucleotide
NBF	neutral buffered formalin
Niflumic acid	3-pyridinecarboxylic acid, 2-{[3-(trifluoromethyl)phenyl]amino}
Nimesulide	N-(4-nitro-2-phenoxyphenyl) methanesulfonamide
NS-398	N-[2-(cyclohexyloxy)-4-nitrophenyl]-methanesulfonamide
PAGE	polyacrylamide gel electrophoresis
PCR	polymerase chain reaction
PD 98059	2'-amino-3'-methoxyflavone ($C_{16}H_{13}NO_3$)
PMS	phenylmethylsulfonyl fluoride
PVD	polyvinylidene fluoride membrane filters
RBC	human red blood cells
RNA	ribonucleic acid

SB 202190	4-(4-fluorophenyl)-2-(4-hydroxyphenyl)-5-(4-pyridyl)1H-imidazole FHPI ($C_{20}H_{14}FN_3O$)
SB 202474	4-ethyl-2(p-methoxyphenyl)-5-(4'-pyridyl)-IH-imidazole ($C_{17}H_{17}N_3O$)
SB 203580	4-(4-fluorophenyl)-2-(4-hydroxyphenyl)-5-(4-pyridyl)1H-imidazole FHPI ($C_{21}H_{16}N_3OSF$)
TBST	Tris buffered saline with Tween 20
TCA	trichloroacetic acid
TRAP	telomeric repeat amplification protocol

IWA Publishing's authorised EU representative for General Product Safety Regulations is Diane D'Arras, 15 rue Duret, 75116 Paris, France, e-mail: safety@iwap.co.uk.

Printed and bound by CPI Group (UK) Ltd, Croydon, CR0 4YY
17/09/2026
02218808-0013